T0291690

# CAMBRIDGE LIBRARY COLLECTION

*Books of enduring scholarly value*

## Physical Sciences

From ancient times, humans have tried to understand the workings of the world around them. The roots of modern physical science go back to the very earliest mechanical devices such as levers and rollers, the mixing of paints and dyes, and the importance of the heavenly bodies in early religious observance and navigation. The physical sciences as we know them today began to emerge as independent academic subjects during the early modern period, in the work of Newton and other 'natural philosophers', and numerous sub-disciplines developed during the centuries that followed. This part of the Cambridge Library Collection is devoted to landmark publications in this area which will be of interest to historians of science concerned with individual scientists, particular discoveries, and advances in scientific method, or with the establishment and development of scientific institutions around the world.

## The Home Life of Sir David Brewster

*The Home Life of Sir David Brewster*, originally published in 1869, records the remarkable life of inventor, physicist, mathematician and astronomer, Sir David Brewster (1781–1868). Written by his daughter, Mrs Margaret M. Gordon, the book is aimed at a non-academic audience, and details the extraordinary life and work of this amazing scientist, who began his studies at Edinburgh University at the age of just twelve, and who is best known for his invention of the kaleidoscope and of the apparatus that initially formed the structure of the core of the lighthouse, and thus his work on the polarization of light. Mrs. Gordon cites Brewster's many activities, including the publication of over 2,000 scientific papers, though she stresses that she has written about her father as the man, and not the scientist. The book will appeal to anyone interested in the life and career of this undoubtedly brilliant Scotsman.

Cambridge University Press has long been a pioneer in the reissuing of out-of-print titles from its own backlist, producing digital reprints of books that are still sought after by scholars and students but could not be reprinted economically using traditional technology. The Cambridge Library Collection extends this activity to a wider range of books which are still of importance to researchers and professionals, either for the source material they contain, or as landmarks in the history of their academic discipline.

Drawing from the world-renowned collections in the Cambridge University Library, and guided by the advice of experts in each subject area, Cambridge University Press is using state-of-the-art scanning machines in its own Printing House to capture the content of each book selected for inclusion. The files are processed to give a consistently clear, crisp image, and the books finished to the high quality standard for which the Press is recognised around the world. The latest print-on-demand technology ensures that the books will remain available indefinitely, and that orders for single or multiple copies can quickly be supplied.

The Cambridge Library Collection will bring back to life books of enduring scholarly value (including out-of-copyright works originally issued by other publishers) across a wide range of disciplines in the humanities and social sciences and in science and technology.

# The Home Life of
# Sir David Brewster

MARGARET MARIA GORDON

CAMBRIDGE
UNIVERSITY PRESS

CAMBRIDGE UNIVERSITY PRESS

Cambridge, New York, Melbourne, Madrid, Cape Town, Singapore,
São Paolo, Delhi, Dubai, Tokyo

Published in the United States of America by Cambridge University Press, New York

www.cambridge.org
Information on this title: www.cambridge.org/9781108014250

© in this compilation Cambridge University Press 2010

This edition first published 1869
This digitally printed version 2010

ISBN 978-1-108-01425-0 Paperback

# THE HOME LIFE

OF

# SIR DAVID BREWSTER.

*Edinburgh : Printed by Thomas Constable,*

FOR

EDMONSTON AND DOUGLAS.

| | |
|---|---|
| LONDON | HAMILTON, ADAMS, AND CO. |
| CAMBRIDGE | MACMILLAN AND CO. |
| GLASGOW | JAMES MACLEHOSE. |

# THE HOME LIFE

OF

# SIR DAVID BREWSTER

BY HIS DAUGHTER

MRS. GORDON

AUTHOR OF " WORK ;" " LADY ELINOR MORDAUNT ; OR SUNBEAMS IN THE CASTLE ;"
" SUNBEAMS IN THE COTTAGE ;" " LEAVES OF HEALING ;" " RIGHTS AND
WRONGS ;" " FASHIONS OF THE PERIOD," ETC. ETC.

EDINBURGH: EDMONSTON AND DOUGLAS

1869.

# PREFACE.

In placing before the public the following notes of my father's life, taken from a home point of view, I am too well aware of the unfavourable criticisms which I almost necessarily incur. I have persevered in the face of many difficulties, however, because a strong wish is known to exist among the unscientific (for whom alone I write) to have a more familiar and accessible record of a useful and brilliant career, than can be expected from the scientific memoirs of Sir David Brewster, which, it is hoped, may soon be undertaken by competent writers. I have not called in his numerous letters, and have principally made use of my own materials, and of what has been placed at my disposal.

The kindness of my father's distinguished colleague, Professor Tait, in revising the allusions to science which have necessarily occurred, secures correctness in this part of my volume. To the other kind friends whose notes appear in the work, or who have gathered information for me, I beg also to return my cordial thanks.

Edinburgh, *Oct.* 1869.

# CONTENTS.

## CHAPTER I.

### THE BIRTHPLACE.

## CHAPTER II.

### THE CHILD.

## CHAPTER III.

### THE COMPANIONS.

## CHAPTER IV.

### THE STUDENT.

## CHAPTER V.

### SETTLING IN LIFE.

## CHAPTER VI.

### NOTES OF LIFE FROM 1810 TO 1814.

## CHAPTER VII.

### NOTES OF LIFE FROM 1814 TO 1824.

## CHAPTER VIII.

### MISS EDGEWORTH—JUNIUS.

## CHAPTER IX.

### NOTES OF LIFE FROM 1824 TO 1830.

# LIST OF ILLUSTRATIONS.

# CHAPTER I.

OH softly, JED, thy silvan currents lead,
Round every hazel copse, and smiling mead,
Where lines of firs the glowing landscape screen,
And crown the heights with tufts of deeper green ;
While mid the cliffs, to crop the flowery thyme,
The shaggy goats with steady footsteps climb.
LEYDEN'S *Scenes of Infancy.*

THE low countries of Scotland combine much of the beauty of the sister kingdoms. We have the sparkling rivers, the purpled hills, the woods of pine and birch, the moory wastes with their varied glories of gorse and heather, while blended with these Scotch attributes, we have cultivated meadows, smiling cottages and hedgerows, that would do honour to the broad lands south of the Tweed. Roxburghshire adds to this twofold beauty the renown of being classic ground, for the wand of the Northern Wizard has given it a world-wide name for all that is interesting in legendary history, and in modern literature. The Tweed, the glorious Border river, is a name which wakes a peculiar thrill of love and interest in all those who have been dwellers on its banks, and the blue Cheviots are scarcely less dear and familiar household words. Various tributaries of the Tweed are likewise dear to Lowland hearts—the Yarrow, the Leader, the Liddel, the Gala, and, very speci-

A

ally to some of us, that clear little stream, the surroundings of which Burns commemorated as

"Eden scenes on crystal JED."

As the poet was in love with a fair Jedburgh lady[1] that
same year, his praises, however, of Jedburgh localities,
must be taken *cum grano salis*. With all deductions, much
true beauty remains in the deep valley, with its wooded
banks and its bright waters, and its varied combinations of colour, for, —blended with the green of summer,
and the spring flushes of orchard blossom, and the rus-
sets of golden autumn,—there are rich-coloured preci-
pices of old red sandstone—called in local parlance
"scaurs,"[2]—which are very noted features in the landscapes of Jed.   On this little river, and in the centre
of its valley, we find the small and not very populous
town of Jedburgh—or Jedworth, as it used to be called,
from *ged*, signifying withes or twigs, and *worth*, a court
or lawn, which suited well its old locality—a town or
court with stretches of forest on every side ;—a noted
and busy place in its day, though its fame was not
always of a pacific or praiseworthy sort.

Only ten miles from the Border, Jedburgh was the
centre of Border warfare.   It was the gathering-place
of Scotish armies, and the favourite point of attack of
the English, who burned it to the ground six or seven
times, while their soldiers frequently occupied the
castle.   It was not a walled town, but its houses were

[1] Miss Isabella Lindsay, daughter of Dr. Lindsay.  She married a Mr.
Armstrong, and became the mother of the late General Armstrong, Master
of the Mint to the Emperor of Russia.  Before Burns left Jedburgh he
presented his portrait to this young lady (1787).—Cunningham's *Life of
Burns.*

[2] Each scaur has its name.  Thus we have Todlaw Brae, Sunny Brae,
Linthaughlee, Hundalee, and even (alas! for poetry) there is the Grumphie
Scaur ! -

so constructed as to admit of but four entrances, which
had gates, and were named Castlegate, Highgate,
Canongate, and Burn Wynd, which circumstance gave
rise to the belief that it was a regularly fortified town.
The Earl of Surrey wrote to Henry VIII. of "the fine
houses, and six good towers" which he destroyed, and
a red flush upon the ruined Abbey still tells its tale of
the fires then kindled. He added, "I found the Scottes
at this time the boldest men and the hottest that ever
I saw in any nation."

The Raid of Redeswire terminated the Border warfare in a manner very creditable to the warlike fame
of Jedburgh. This celebrated engagement took place
in 1575, between the Scots, headed by the Rutherfurds
of Hunthill and Hundalee, and the English, under Sir
George Heron, Keeper of Tyndale. It was fought on
the slopes of the Carter Fell, at the pass into England
by Redesdale, over which marched the invading armies
of both countries. A sudden onslaught of Jedburgh
citizens gave the victory to the Scots, when

> "They raised the slogan with ane shout,
> Fy, Tyndaill, to it—Jedburgh's out!"[1]

In the vicinity of Jedburgh there still remain many
landmarks of the patriotic past. A mile and a half
from the town, on the top of a scaur overlooking the
Jed, are the remains of the celebrated camp of that
"good Sir James of Douglas" who

> "Quhen he was blyth he was lufly,
> And meyk and sweyt in cumpany;
> But quha in battaill mycht him se,
> Anither countenance had he."[2]

It is said that he defeated ten thousand Englishmen

[1] Scott's *Border Minstrelsy*.            [2] Barbour's *Bruce*.

near the Lintalee camp, a victory which " added another
blazon to the shield of Douglas," and secured all the
south of Scotland for King Robert the Bruce, who
was then absent in Ireland. Half a mile above the
camp is Fernyhurst Castle, the ancient seat of the
Kerrs, and the scene of fierce struggles and wild
legends. Then there is Bairnkin, an old fastness of the
Oliphants, and four miles further up the river the ruined
church of Old Jedworth, founded by Ecgred, Bishop of
Lindisfarne, about the year 800. On every side are
traces of peels and camps, especially of the latter, at
Scraesburgh, Gilliestongues, Rink or Camp-town, and
Chesters ; while the frequent caves in the scaurs, as at
Hundalee, Lintalee, and Mossburnsford, were probably
favourite places of refuge for border heroes in their
ever changing tides of fortune. It was at the church
of Southdean,[1] or " Sooden" as it is called, on the Jed,
that the Scotish army, as described by Froissart, assembled for the expedition which terminated in the battle
of Otterburn in " Chevy Chase."

The ballad lore, and the numerous remains of peel-
houses and fortified towers throughout Roxburghshire,
as well as stern historical facts, leave no room for doubt
that the little forest town was frequently the scene of
other encounters and adventures less patriotic and creditable than struggles against the usurpation of England. " Dalesmen " and " marchers," or in other words
thieves and freebooters, carried on a guerilla warfare
from the fastnesses of the neighbouring forests, so that
Jedburgh was kept in a constant moil of civil strife
of a vexatious and ignoble kind, from the earliest
monarchies till the days of the Regent Moray, who

[1] White's *Battle of Otterburn*, p. 23.

exerted his authority with some success against the rule
of rapine which existed in the Border county.   The
proverbial expression of " Jeddart justice " itself tells a
vivid tale of the Lynch-like manners and customs of
the place, where it was said to be the fashion " to hang
first and try afterwards !"[1]   A ferocious-looking weapon
of Border warfare, a sort of battle-axe, helped on the
dubious character of that anti-pacific locality by its
satiric name of " a Jedburgh staff"!

Jedburgh Castle was entirely demolished as early as
1490.   It was situated on a hill at the head of the town,
and was a favourite residence of the early Scotish kings.
David I. held his court principally at Roxburgh ; but had
also his royal castle at Jedworth ; Malcolm IV. died
within its walls ; Alexander III. celebrated there with
great festivities his marriage with the beautiful Jolande
of France.   Mr. Cosmo Innes gives the following interesting account of Jedburgh at this period :—" Alexander III. and his queen, the daughter of the lordly
De Coucy, chose Jedburgh and its lovely valley as a
favourite residence.   After the death of that king, John
Cumin rendered his account as bailiff of the king's
manor of Jedworth, in which he charges himself with
66s. 8d. as the rent of the new park, which used to be
the place of the queen's stud (*equicium reginæ*), 26s. 8d.
for the sales of dead wood ; and states his outlay for
mowing 66 acres of meadow, and for winning and carry-

[1] The magistrates of Jedburgh had the power of life and death as well
as the barons, and their arbitrary exercise of it at a comparatively recent
period reminds one vividly of the old Jedburgh proverb.   In 1715—" Mar's
year," as it was called—some suspected rebels were brought before the
zealous magistrates.   Only one crime could be clearly "proven," which
was that they were "real natural-born Irishers," on which conviction they
were forthwith taken to "the goose pool, and hadden doun, and drownit
till they were dead."

ing it for forage for the castle. *Item*, for 900 perches
of ditch and hedge (*fosse et haye*) constructed about
both the wood and the meadows of Jedworth, 116s. 6d.
I think I cannot be mistaken in translating these words
ditch and hedge; and if so, you have by far the earliest
instance of such a fence on record. I suppose the
wood so enclosed may have been the bank of Fernyhurst, and the meadows those fairy fields by the side of
the Jed which form one of the most beautiful and
peculiarly Scotch scenes I have ever seen."[1]

Queen Mary Stuart kept court at Jedburgh on several
occasions, her residence being an ancient house, carefully preserved, and associated with her name. From
thence too she started on her wild visit to Bothwell in
1566, at Hermitage Castle, in Liddesdale, twenty statute
miles from Jedburgh, a distance which she performed
to and fro in one day on horseback, on the roughest
roads, and in the midst of Border perils. The fatigue
and anxiety so shook her nerves that she lay sick of a
lingering fever for a whole month, in the dark wain-
scoted rooms of the old Jedburgh house.

Jedburgh was also a favourite resort of the ecclesiastical world. Its noble Abbey—presenting in its chancel,
a fine specimen of the earlier Norman architecture, as
well as of the later in the beautiful door of the cloister,
and in the western gable—was founded by David I.,
and owes its demolition, not to the Reformation, but to
English soldiers. There were also large monastic and
conventual establishments, of which but the names
remain, such as the "Friars'" and the "Lady's Yards."
The gardens and the orchards, which so beautifully
mingle with the town buildings, owe their celebrity,

[1] *Scotland in the Middle Ages*, by Cosmo Innes, pp. 125-6.

it is said, to the careful culture of the old monks. Jedburgh pears have long been widely famed, and the Jedburgh plum is of rare excellence, rivalling the green gage. Magnificent old pear-trees from French grafts, planted by the monks, were much admired for their picturesque beauty, many being three feet in diameter; but those that remain, though old and gnarled, are but their offshoots, and none of them date further back than three or four hundred years. Some of the ecclesiastics of Jedburgh, however, had wider celebrity than that of being good horticulturists, and Adam Bell, one of its Carmelite friars, wrote a History of Scotland.

At the foot of the Canongate, spanning the Jed, there is a fine old bridge of ashlar, with three circular ribbed arches, dating from the time of the building of the Abbey. In "the '45,"[1] when "Prince Charlie" entered Jedburgh by this bridge, it was the scene of a vehement and rejoicing Jacobite welcome. He took up his quarters at the hostelry of The Nag's Head, still in existence.

After all the sore, yet, on the whole, successful battles which Jedburgh waged for the liberty and independence of Scotland, it was at last the Union with England, which, by ruining its trade, did it more harm, and brought it nearer absolute ruin, than Surrey and all his soldiers. In the middle and towards the end of last century Jedburgh had greatly declined; its citizens were called "idlers," the old trees were cut down, the churchyard lay open to all intruders, and the fine old Abbey was a quarry from whence was taken the material for parochial buildings. From one of its aisles

[1] Mr. Veitch's mother remembered seeing on this occasion, from Bonjedward, the Highlanders shoot the sheep on the opposite slopes of the hill of Ulston.

ascended daily the hum of the lessons or the shouts
of the play-time of unruly schoolboys. Jedburgh, in
short, was in a bad case, and, following the general rule
in such circumstances, Superstition abode within its
shades. Witches and fairies supplied the place of more
substantial inhabitants. The Witches' Green was a real
and abiding locality. A window in the Abbey was
called the Witch's Wheel, through which it was believed that veritable fairies danced by moonlight.

Neither the presence of superstition, desolation, and
ruined stone and mortar, nor the absence of red chim-
neys, successful money-making, and the modern wear
and tear of life, could prevent a new and deeper
*renaissance* of the old energy and activity of Jedburgh.
It was not only that men of poetry and adventure
began to arise throughout Roxburghshire, such as
Leyden, Thomson,[1] Pringle,[2] and others ; such gifts
might be expected amidst the stirring memories
of the past, and the beautiful scenery of the present.
The deeper gifts of intellect, however, appear about
this time to have been lavished with no scanty hand
upon the men of Jedburgh, so that the peculiarly gifted
ones who were yet to arise in their midst found a
fit nursing-place, and helping hands on every side to
cherish and uphold. The thews and the sinews of the
men of old reappeared in the mental conformation of
their descendants. Battles of a higher life went bravely

[1] One of the poems of Thomson, which is less known than his *Seasons*,
is an *Ode to Sir Isaac Newton*—

" When Newton rose, our philosophic sun,"

so that appreciation of science was not wanting even among the poets.

[2] Thomas Pringle, author of *Scenes of Teviotdale, Ephemerides,* and a
series of *African Poetical Sketches,* of which " The Bush Boy " is the best
known. He was connected with *Blackwood's Magazine* at its commencement.

on in the fields, forests, streets, and ruins, where in the
old-world days valiant yet useless conflicts of flesh and
blood had been lost and won. Conquerors went forth
conquering from the valley of the Jed, subduing ignorance, combating superstition, inventing new blessings
for the nations, discovering new laws and beneficent
designs, and bringing glory to Him who reigneth over
mind as well as matter.

## CHAPTER II.

### THE CHILD.

> By what astrology of fear or hope
> Dare I to cast thy horoscope !
> Like the new moon thy life appears
> A little strip of silver light,
> And, widening outward into night,
> The shadowy disk of future years !
> And yet, upon its outer rim,
> A luminous circle faint and dim,
> And scarcely visible to us here,
> Rounds and completes the perfect sphere.
> A prophecy and intimation,
> A pale and feeble adumbration,
> Of the great world of light that lies
> Beyond all human destinies !
>                                 LONGFELLOW.

> Being not propped by ancestry (whose grace
> Chalks successors their way), neither allied
> To eminent assistants, but spider-like,
> Out of his self-drawing web, he gives us note
> The force of his own merit, makes his way ;
> A gift that Heaven gives for him.
>                             SHAKESPEARE.

IN the Canongate of Jedburgh, where it slopes down to the river and the old bridge, there is a plain substantial house,[1] which is frequently pointed out to visitors.    In a westward room of this, his father's house, a little child drew the first breath of a life of eighty-six years on the 11th of December 1781, a little after mid-day.   DAVID BREWSTER was the third child and second son of James Brewster, rector of the Grammar School of Jedburgh, a man of sterling worth, and

[1] This house is now a model lodging-house, having been bought by the Marquis of Lothian, and fitted up for that purpose.

accounted as one of the best classical scholars and
teachers of his day. After completing his course at
the University of Aberdeen, Mr. Brewster "gained a
doctorship in the Grammar School of Dundee, merely
by superior merit, in a comparative trial of half-a-dozen
candidates."[1] In 1771 he was chosen by a large majority to be master of Jedburgh school. It had been
founded and endowed by Bishop Turnbull, the founder
of Glasgow University, and its endowment bestowed
the title of rector on the schoolmaster, as is still the
case in a few other Scotch towns. It was, however,
also the parish school, and had long been one of high
repute, several remarkable men having received their
education there,—amongst others, at the time of the
Reformation, John Rutherfurd, Principal of St. Salvator's College, St. Andrews ; in the following century,
Samuel Rutherford, the Covenanter, who was born at
Nisbet, four miles from Jedburgh ; and in more recent times, Thomson, the poet of the Seasons, who
was born at Ednam, but whose father, when he was
only two years old, was translated to Southdean on
the Jed. Mr. Brewster, in whose hands the school
suffered no loss of good fame, was considered a stern
disciplinarian, and the common saying was, that he "was
the best Latin scholar, and the quickest temper in Scotland;" yet the sterner mood did not always prevail.
On one occasion, William Somerville[2] and a companion
had been ordered to retire to the library, an adjoining
apartment, for purposes of serious punishment. It was,
however, a fine "hunting morning"—the fox-hounds were

[1] Extract from a letter, highly commendatory of Mr. Brewster, from one
of the ministers of Dundee to the Rev. Dr. James Macknight, the celebrated
Biblical critic and commentator, then minister of Jedburgh.
[2] Afterwards the husband of the celebrated Mrs. Somerville.

passing—the temptation was irresistible—they scrambled out of the window, and enjoyed a day's sport instead of a flogging. The rector's countenance the next day was regarded with fear and trembling—yet he was in a good humour, and had apparently forgotten the empty room and the open window. He had the exceedingly unpopular fault in a schoolmaster, of occasionally forgetting the customary dispensation of holidays, being absorbed in his own love of study. On one occasion the shortening days and the ripe yellow harvest telling a tale of the passing away of the proper holiday time, the rector was surprised by a sheaf of ripe ears waving on his desk, placed there by a young Armstrong[1] and a Lindsay. The rector proved "quick at the uptak," the hint was understood, and the holidays at once proclaimed. He possessed, with all his sternness, much attractiveness of person and manner, more, it was always said, than was inherited by any of his family. David, however, resembled his father in expression and contour, and, though without his regular features, was an exceedingly beautiful child.[2]

Mr. Brewster resigned his charge on account of age and infirmities in 1803—a resignation which was received with great regret, a retiring salary being voted to him by the heritors as a marked tribute of respect. He died very suddenly in 1815, when upwards of eighty, in the beautiful Manse of Craig, the home of his eldest son, much beloved and regretted by those who in the contemplation of his holy life, patient blindness, and most gentle Christianity, had long forgotten the severities of his home and school training. He rests

[1] Afterwards General Armstrong, Master of the Russian Mint.
[2] So much so that his old nurse used to look at him and say, "Eh, laddie, ye'll mak many a lassie sigh and set aside her supper!"

in "sure and certain hope" in the picturesque and
rocky churchyard of St. Skeoch in Forfarshire. An
amusing instance of the respect in which Mr. Brewster
was held in Jedburgh is still told. An old man bought
a frying-pan at the sale of the household effects which
took place in 1806, when the family left Jedburgh,
which frying-pan became henceforth a sacred relic in
his eyes. All attempts to purchase it were indignantly
negatived, and he treasured it in memory of "Sir
Daavid's fayther," till his death, in extreme old age, a
few months after that of the subject of this Memoir.

David was only nine years old when his mother died,
but ever spoke of her with that tender deference which
great men peculiarly feel for their mothers. Margaret
Key was the daughter of "James Key, Junior" and
"Grisel Scott," who were married in Dundee about
1745. The Keys and the Scotts, who intermarried during several generations, were burgesses and shipmasters
of Dundee. One son of this marriage was Dr. Patrick
Key, a favourite uncle of David's, who practised as a
medical man in Forfar, with much ability and repute
in his profession, the arduous duties of which exhausted
his constitution, and he died at a comparatively early
age, much mourned by the Brewsters.[1] Margaret was
the only daughter, and married Mr. Brewster of Jedburgh in 1775.

Mrs. Brewster is said to have been a very accomplished woman for the time she lived in; but some
delicate broidered work, in which she especially ex-
celled, a small book of MS. religious extracts, and the
knowledge of her exceeding fragility, which carried her

---

[1] Dr. Key married Anne Binny, daughter of Mr. Binny of Newmill, a
family of staunch Jacobite Episcopalians. John Binny Key, Esq., and
Thomas Key, Esq., are sons of this marriage.

off soon after the birth of her youngest child, at the age of thirty-seven, are all that remain to us of her. A stone in the old Abbey churchyard records successive bereavements thus :—" In memory of Margaret Key, the beloved wife of James Brewster, Rector of the Grammar School of Jedburgh, who died on the 1st of September 1790, and of Jane Gordon, their infant daughter, who died on the 30th of May 1790, aged one year." For some little time Mr. Brewster's stepmother, of the same age as himself, a woman of considerable beauty and much kindliness, was a blessing to the motherless family, but she too was soon removed, and Grisel, the only sister, three years older than David, from thenceforth filled a mother's place. She possessed much of the family talent, and was not slow in discovering the genius of her second brother. Her appreciation of it, and some natural over-indulgence, contrasted unfavourably with the severity of paternal rule. When David was but a little fellow he used to punish this indulgent elder sister for any unwonted stretch of authority, by turning the contents of her chest of drawers on the floor in confusion dire, knowing that her ruling passion was love of order. Whatever the juvenile offences of the somewhat spoiled child might be, they never occurred in connexion with learning. It is recorded that though David was never seen to pore over his books like other schoolboys, yet by some mysterious process he always " had his lessons" notwithstanding ; he kept a prominent place in his classes, and was frequently applied to for assistance by his school-fellows. It was in these days of childhood that a dilapidated pane of glass in an upper window of his father's house produced the inquiring thoughts which led him afterwards to search into the mysteries of refracted light. David was however no *rara avis* in the rector's household. Intellect and learning were the paternal inheritance shared by all the brothers. James and George possessed excellent abilities, while Patrick, the youngest, was considered to possess a higher portion of genius and intellect than his second brother. But David had received that other and better gift than genius, the power of making use of whatever he possessed.

The Brewster family, though nurtured under the lowly roof of a Jeddart schoolmaster, were not wanting in tales and legends of the past. The step-grandmother, probably gifted with more kindliness than good sense, used, we are told, "to relate by the hour tales of their ancestors," which the youthful mistress of the house eagerly devoured, and related again to her brothers in later life. Sometimes these mythical claims have greater effect upon the imagination than those which are recognised and undeniable ; but nursery tales had no effect upon David Brewster, who was thoroughly " a self-made man." It would be wrong to say that he rejoiced in being so, for the thought of it one way or another never seemed to enter into his mind. He was entirely unconscious, ignorant, and indifferent as to all genealogical failures or possessions—an ignorance and an indifference which he communicated to all those within his immediate influence.[1] There have not been

[1] A lady genealogist having questioned my father on this subject in his later years, he told her that while by his mother's side he was entirely Scotch, he was not so on his father's, as the Brewster family had, he believed, come from England several generations before. He added characteristically, "The books say that I come from a branch of the Brewsters of Wrentham, but I neither know, nor do I care." He might have quoted Sydney Smith's favourite saying of Junot's, " Je n'en sais rien, moi, je suis un ancêtre."

wanting those in more recent times who have thought
it ought to be known that the nursery tales were not
altogether mythical, and that had Brewster dabbled in
Heralds' Offices and their occult lore, he might have
proved that he was descended from a branch of an old
English family, whose name was undoubtedly Saxon—
the Brewsters of Wrentham in Suffolk, descendants of
different branches of which frequently claimed kindred
with him.[1]  The only hint of the kind which was not
received with the greatest languor and indifference was
significant of his love of liberty and religious toleration,
—it was that which suggested his descent from William
Elder Brewster, the Puritan postmaster, printer, and
publisher, who led the noble band of English "Dis-
senters" in the "Mayflower" from England to America
in 1610, whose interesting Memoir was sent to him by
some of the Brewster family long settled in New England, where the Pilgrim Fathers sought and found
"Freedom to worship God."

The Grammar School had been held in the Abbey for
a considerable time, in what was in consequence called
the "Latin aisle," but the present school-house, in which
David Brewster was educated, was built in 1779.  The
old aisle was, however, still part of the playground of
the boys, and received the name of the "Howff," an
old Scotch word signifying a place of relaxation, and
also of refection, for the boys used to eat their mid-day
meals there, preparatory to more active diversions.
Many were their exploits in dangerous climbing
amongst the ruined arches and up the lofty towers,

[1] Of these Cardinal Brewster, Esq. of Greenstead Hall, Essex, representative of the Brewsters of Hedingham, and his father, the late James
Brewster, Esq. of Ashford Lodge, were personal friends of my father, who
much appreciated their pleasant home circles.

stepping across the rent or "gap" at a height of ninety
feet. The choice attraction was, boy-like, the mere
danger, with also an eye to the nests of owls, jackdaws,
swifts, and such like, in all which exploits David and his
brothers sought to excel. A more ghostly charm attached
itself to one of the vaults under the town steeple, however, in which was kept the gibbet. To enter this vault
in the dark night and touch the horrible object, was
considered by the boys to be the height of human daring.
David used to record in after years the *one* time when
he accomplished this feat, and fled immediately after,
feeling as if he were pursued by the spectres of all the
hanged men of Jedburgh. "Hangie," as the boys
dubbed the functionary who came occasionally to fulfil the last penalty of the law (the last permanent
hangman was paid off in 1666), was also an object
of intense fear and attraction. David described the
irresistible impulse which led him and his companions
to throw stones at the door of "Hangie's" temporary
abode, and their horror when he used suddenly to
emerge and give them chase.

The love and fear of the superstitious which had so
long reigned in Jedburgh came nearer home to David
than even these localities. Behind his father's house
was a little cottage, of which only the gable now
remains. It was shaded by a favourite apple-tree,
and was the dwelling of the old nurse of David's childhood. She was still the repository of the Brewster
children's hopes, joys, and sorrows, and the favourite
employment of a winter evening was to spend it
with her. So many were the stories of ghosts and
goblins with which she garnished the evening's entertainment, that it generally ended in the poetical jus-

tice of the old woman having to leave her easy-chair and her cosy fire, to convoy the frightened children across the garden, with her protecting apron thrown over their heads. David's recollections of the rosy apple-tree of summer, and the fascinating " bogle" stories of winter, were very vivid ; but he also related how the usual effects of such a training resulted in his suffering from superstitious fears even up to the mature years of manhood.

## CHAPTER III.

### THE COMPANIONS.

SAY, shall my hand with pious love restore
The faint fair picture Time beholds no more,
How the grave senior  .  .  .  .
Saw from on high, with half paternal joy,
Some spark of promise in the studious boy,
And bade him enter with paternal tone
The homely precincts which he called his own.
    Yet in each leaf of yon o'ershadowing tree,
I read a legend that was traced by thee,
Yet in each line that scores the grassy sod
I see the pathway where thy feet have trod ;
Though from the scene that hears my faltering lay
The few that loved thee long have passed away,
Thy sacred presence all the landscape fills,
Its groves and plains and adamantine hills.
                    OLIVER WENDELL HOLMES.

IT is always interesting to know the quantity and
quality of the mental pabulum which has fed and
nourished the men who have left their mark on the
age.   In David Brewster's case the quantity was great,
and the quality of unusual excellence.   Besides several
intelligent gentlemen resident in the neighbourhood,
there were among the citizens of Jedburgh, during
Brewster's youth, various men of original character,
scientific tendencies, and especially of inventive genius.
There was Mr. John Ainslie, land-surveyor, well known
in Edinburgh for his large map of Scotland ; Mr. Alexander Scott of Fala, afterwards factor to the Earl of
Hopetoun, a man of sterling integrity of action and
speculative mind, inquisitive about every new inven-

tion and discovery, whether steam-engines or balloons, reaping machines or orreries; Mr. George Forrest, gunmaker, whose inventions were honourably mentioned in the *Edinburgh Philosophical Journal;* Mr. William Hope, ironmonger, who invented a printing-press, for which he held a patent for many years, and who also introduced gas into his shop and foundry at a very early period; and Mr. Gibson, watchmaker, who made barometers and reflecting telescopes. We are told also "that there then resided in the town several families, if not wealthy yet of good intelligence and respectability; amongst whom were the Rutherfurds; the Murrays, cousins of the Earl of Home; the Halls, of the Covenanting family of Haughhead; the Fairs and the Shortreeds. The Halls were of the medical profession; and Mrs. Hall, the relict of one of the brothers, latterly living in London, was a contributor to the principal periodicals, and was honourably mentioned in the obituary of the *Athenæum* not many years ago; she wrote an interesting article on Jedburgh in the time of Boston for one of the early numbers of *Fraser.* James Fair, one of Brewster's companions, though a military man, aspired to literature, both in prose and poetry. Sheriff Shortreed is well known as the early friend of Sir Walter Scott, and his son, Major-General Shortreed, has distinguished himself by his mathematical and scientific attainments as an engineer in India. Amongst the contemporaries of Brewster may also be mentioned Robert Easton, known as 'Lang Rab'—a land measurer, fond of astronomical and botanical pursuits, whose apology for blunders in his calculations was— 'All men err since Adam fell!'"

Dr. Somerville, the minister of the parish, and the

successor of Dr. Macknight, whom Burns described as
" a man and a gentleman, but sadly addicted to pun-
ning," was the author of several respectable literary
works; he was the historian of William and Mary,
and Queen Anne, and wrote a volume of *Reminiscences
of his own Times.* He was a man of great literary
research and industry, and he was the kind friend
and patron of the young men who in the course of
his unusually long life and ministry crowded around
him, before entering on their several busy lives. He
availed himself of their services as amanuensis in his
literary labours, explaining to them the art of composition, the nature of idioms, and the reasons for preferring
one form of speech to another. Brewster had his turn in
the office of amanuensis, and often recurred with much
interest to the hours thus profitably spent. It was
a great disappointment to him when he found, long years
afterwards, that the well-remembered room in which
this congenial work was carried on had perished with
the old manse. He continued occasional correspondence, in a tone of much respect and regard, with the venerable Doctor, who wrote the article JEDBURGH for the
*Edinburgh Encyclopædia.* Dr. Somerville died at the age
of ninety, in 1830. It is interesting to trace the effects
of these early lessons, not only in Brewster's peculiarly
careful composition, but in the zest and interest with
which he would point out to those around him, during
the correction of proofs, the why and the wherefore
of the minutest alteration, and in the peculiar quickness of his long-practised ear for the formation and
sound of language, though utterly destitute of musical
appreciation. Dr. Somerville was the uncle and father-
in-law of Miss Mary Fairfax (Mrs. Greig), who married

her cousin and second husband, Dr. William Somerville, in 1812. Owing to the absence of her father, Sir William Fairfax, on public service, she was born and nursed in the manse (one year later than David Brewster), and was looked on by her uncle and aunt as if she were their own child. She still lives to show to the world what woman can accomplish of intellectual and scientific work, without sacrificing one iota of her feminine and household gracefulness and dignity. The literary character of Dr. Somerville, his family and connexions, brought to Jedburgh many visitors of intellectual tastes and superior habits, whom it was both pleasant and advantageous for young men to meet.

But of all the helps and training, the greatest that Brewster met with was from James Veitch of Inchbonny, ten years his senior, a man so remarkable himself for genius and talent, that it is passing strange so little is known of him in this generation.

Half a mile to the south of Jedburgh, on the Newcastle road, in a charming little valley, environed by lofty banks of wood and red rocky scaurs, is situated a substantial dwelling, most pleasant to sight and sound, with its walls covered with pear-trees, its sunny little garden, its hives of bees, its song-birds, and its murmuring brook. There resided James Veitch, one of a respectable family who had possessed that beautiful little property since 1730, although obliged to combine some manual labour with their inheritance. The sort they chose seems to have been known in the family for good three hundred years, as in the neighbouring churchyard of Bedrule there is an ancient burying-place of Veitches, who had for their monumental emblems an axe and hand-saw. Dr. William Veitch of Edinburgh,

the well-known "Grecian," is descended from the family
buried at Bedrule, so that in this case, as in that of the
Inchbonny branch, manual labour and mental gifts
seem to have gone hand in hand.

James Veitch, to the ordinary education of his class
at that time, had added an amount of self-education
which would have qualified him for any situation in
life.    Had he fully known his powers, been actuated
by ambition, or followed the advice of friends such as
Sir Walter Scott, he might have risen to distinction in
other spheres of greater publicity.    He wished for
nothing better, however, than to throw the subtle halo
of genius even upon his humble daily occupation, while
his choicest relaxation was found in abstruse study.

Originally taught the making of ploughs by Small,
Veitch improved them by lightening them, and reliev-
ing the draught, both as to the form of the mould-board
and the bearing of the beam.    He contributed several
articles on these improvements, illustrating mechanical science, to the *Edinburgh Encyclopædia*.    There
was scarcely any work of mechanical skill, however,
which he was unequal to undertake, and he executed
many delicate pieces of work usually done only in
Edinburgh and London.    Whatever he did was done
not in mere imitation, but on scientific principle, calculation, and experiment.    The construction of telescopes.
was his most favourite occupation.    The curves of his
specula, and also of the lenses for achromatic object-
glasses, he determined most carefully and laboriously.
Much of his time was, however, wasted in the mere
mechanical work, in which he delighted, of tubes,
stands, and other apparatus, which could have been
better done by ordinary workmen.    The timepiece

engaged his attention as well as the telescope, and, with the late Earl of Minto, manifold were his measurements of heights and distances, the use of the barometer being carefully tested by the circle or sextant. He was extremely fond of calculation, and devoted much time to finding the places of the planets, the eclipses of sun and moon, occultations of stars, the transit of Mercury and Jupiter's satellites, often unnecessarily, as he might have found the same from the Nautical Almanac, accommodating the projection to the latitude and longitude of the place. He was the first to discover the great comet of 1811, as is mentioned in the fourth edition of Mrs. Somerville's *Connexion of the Physical Sciences.* In the years 1827-8 he was engaged in ascertaining the relations of the old Scotch local standards of weights and measures to the Imperial, as established by Act 5 Geo. IV. cap. 74, and drew up the reports for the counties of Roxburgh and Selkirk, acting in that capacity for Berwickshire along with Mr. James Jardine, C.E. He drew up and published tables for each of the three counties for the conversion of the old measures into the new. The assize at Selkirk was presided over by Sir Walter Scott. On his way to Selkirk to make the experiments on which the report was founded, Veitch called at Abbotsford for instructions, and found Sir Walter in his study, who received him with his usual kindness, but manifestly was quite a stranger to the subject on which he was soon to adjudicate as sheriff; but when the day of the assize came, he showed that he had meanwhile thoroughly studied it, and his charge to the jury was brilliant and effective. Veitch was also among the first to observe several other comets, and one especially, of great brilliancy, which appeared only thrice, under cloudy weather, in the morning, and seemed to have escaped the attention of astronomers. He assisted Mr. Francis Baily in the observations made at Inchbonny of the great annular eclipse of the sun in May 1836, of which an account is given in the *London Astronomical Transactions.* He was extremely interested, not only in the construction, but in the management of the microscope ; in fact, there was scarcely any branch of learning and research to which this remarkable man of industry and science did not turn his attention. He was not without honour even in his own country, and he was much appreciated by many at a distance. He had a kind invitation from Sir William Herschel, who had frequently heard of him from mutual friends, to visit him at Slough, and see his instruments and operations. He made the acquaintance of Dr. Wollaston, when that philosopher was on a visit to Jedburgh with Mrs. Somerville, and of Professor Sedgwick in later years, while his correspondence with Professor Schumacher, Professor Playfair, Sir Thomas Brisbane, Lord Minto, and others, shows how fully able he was to hold scientific intercourse. Sir Walter Scott was a firm and valued friend ; he used to say, " Well, James, when are you coming amongst us in Edinburgh, to take your place with our philosophers?" and the reply generally was, " I will think of that, Sir Walter, when you become a Lord of Session." To Mr. Ellis of Otterburn Sir Walter writes in April 1818 :—" I heard these particulars from James Veitch, a very remarkable man, a self-taught philosopher, astronomer, and mathematician, residing at Inchbonny, and certainly one of the most extraordinary persons I ever knew. He is a connexion of Ringan Oliver, and is in possession of his sword, a very fine weapon. James Veitch is one of the very best makers of telescopes, and all optical and philosophical instruments, now living, but prefers working at his own business as a plough-wright, excepting at vacant hours. If you cross the Border, you must see him as one of our curiosities ; and the quiet, simple, unpretending manners of a man who has, by dint of private and unaided study, made himself intimate with the abstruse sciences of astronomy and mathematics, are as edifying as the observation of his genius is interesting."[1]

Sir Walter sent an artist to take a portrait of his old uncle, who resided near Jedburgh, and also one of James Veitch, with the latter of which he wrote to his friend Mr. Shortreed that he was much pleased. He had also a clock made by Veitch, for which he prepared a place, and wrote :—" As I am about to build at Abbotsford, I will not trouble you to fetch over the clock till that job is finished ; I will then have a better and more distinguished situation for the work of your hands. We will talk over this when I come to the Circuit." At the pass between Jedburgh and Inchbonny, on Veitch's property, is the celebrated precipice where the red sandstone in horizontal beds covers the vertical grauwacke, separated by layers of conglomerate, in a striking formation, which attracted Hutton's attention, and which he has rendered classical by giving a drawing and interesting description of it in his *Theory of the Earth.* It became, in consequence, a place of great attraction to the geologists of that time.

[1] Letter to James Ellis, Esq., printed in Willis's *Current Notes,* Jan. 25, 1856.

The Circuit Court brought regularly, in spring and autumn, a visit of the judges and advocates, under the guidance of Sir Walter Scott, to see this geological lion, as well as the observatory and philosopher of Inchbonny. On one occasion Lord Jeffrey, on coming down the dark stair from seeing the planets, was heard repeating the verse of the Scotch metrical psalm—

"Yea, though I walk in death's dark vale,
Yet will I fear none ill."

Sir Walter Scott never omitted at these visits to hold in his hand the sword of Ringan Oliver, or properly Oliphant,[1] "a certain bold yeoman, the strongest man in our country," who dwelt in old times at the Tower of Smailcleugh, three miles from Jedburgh, and was a hero after Sir Walter's own heart, although he fought on the side of the Whigs at the battle of Bothwell Bridge, with the said good sword, which is still carefully preserved.

A little to the north of the present house of Inchbonny there stood, towards the end of the last century, a workshop, of which, unfortunately, not a stone or beam remains; from it went forth good and useful ploughs to the agricultural world, always constructed, however, as we have seen, on scientific principles, showing forth the inventive genius of their maker.

It is not to be wondered at that this scientific workshop on the Jedburgh turnpike became a gathering-place for all the young men of intelligence in the neighbourhood, most of them being in training for the ministry, for medicine, and other liberal pursuits. They had lessons in mathematics and mechanics, and in natural science generally, but especially in the

[1] Mr. Veitch's great-grandmother was Ringan Oliphant's sister.

favourite science of astronomy.    The telescopes were
tested in the day-time, by the eyes of the birds perch-
ing on the topmost branches of the " King of the
Wood," a noble relic of the past forest days,[1] about
half a mile from Inchbonny.    When the bright sparkle
of the bird's eye was distinctly visible by day, James
Veitch's specula and lenses were considered fit to show
the glories of the sky by night.    Nor were discussions
in theology wanting.    Mr. Veitch was a truly God-
fearing man of the old Scotch school, and was well
fitted to guide such conversation.    One of a later generation,[2] receiving good advices from the then aged man,
recalls vividly the energy and pith of Veitch's concluding words, " Ye hae a cunning adversary, mind that;
Satan 's no a 'prentice hand !"

Many original characters swelled the group.    Mr.
James Scott, son of the Relief minister, was an object
of great interest and amusement, from his cleverness
and eccentricity ; of peculiar appearance too, being
afterwards described as " a little man, of long corpulent body, short legs and large head, with a brown
wig and wide hat."    Although at first not appreciated
in Edinburgh, he became a very popular preacher,
attracting crowds to the large chapel in the Cowgate,
now occupied by Roman Catholics, though hints were
rife in Jedburgh that certain volumes of sermons
borrowed from the Inchbonny bookshelves were returned in a suspiciously well-thumbed state at the

[1] Another magnificent tree near Inchbonny is called the Capon Tree,—
one of the many traditions of its name being a " kain" or tax of as many
fat capons as could shelter beneath its far-spreading branches, paid to the
monks by the tillers of the Abbey lands.

[2] The Rev. Mr. Ritchie, the present incumbent of Jedburgh, who keeps
up the character of his predecessors for intelligence and urbanity.

very texts of the most popular of his discourses.
David afterwards expressed much regret that he had
never heard him preach, so vivid were his recollections
of the picturesque little man and his odd ways. As
might have been expected, practical scientific jokes
went merrily on among the young Inchbonny philosophers. An electrifying-machine was an unfailing
source of interest, and we are told that " a favourite
experiment of the youngsters was that of artfully placing the conducting-wires on the seat usually occupied
by a frequent visitant of dignified presence, Laird Gray,
with his broad blue bonnet and white locks, and worsted
overalls reaching above the knee. Delighted to get
him seated, they sedulously engaged him in conversation until one or two in a separate apartment worked
the machine to its full discharge, when the Laird's
sudden jump and tone of amazement was the catastrophe which called forth their warm sympathy and
anxious philosophic investigation into the cause, the
Laird being satisfied that the mischief lay in a sparkling piece of quartz, which no persuasion would induce
him to approach. The fun was at sundry intervals repeated, and they always took care to have the stone in
suspicious juxtaposition."

The charm of foreign visitors was not lacking. The
workshop was visited daily by the French prisoners
who resided in Jedburgh on parole during the French
war. Several naval officers were peculiarly noted for
their zeal in science. " M. Charles Jehenne,—captured
at Trafalgar, who from the mast-head observed Nelson's
fleet bearing down on the French : ' They saw us,' he
said, ' before we saw them,'—successfully constructed
a telescope. Another old naval lieutenant, M. Scot,

with a long grey coat, was to be seen with every gleam of sunshine at the meridian line, with compasses in hand, resolved to determine the problem of finding the longitude."

Such was the man to whom Brewster was attached from his earliest youth, and such were the companions and the scene of his truest and best education. I have driven slowly past the site of that little workshop, some seventy years after those happy days, while, with the peculiar tearful light which always came into his eyes when warmly touched, he pointed out the localities with the freshest, liveliest interest. He was the very youngest of the quaint and varied group. When he began his visits I do not know, but we find that at the age of ten he finished the construction of a telescope at Inchbonny, which had engaged his attention at a very early period, and at which he worked indefatigably, visiting the workshop daily, and often remaining till the dark hours of midnight, to see the starry wonders and test the powers of the telescopes they had been making.[1] His brothers were often with him, but they had other outlets of amusement, while David's ardent love of science made him prefer the hours of study and observation at Inchbonny to all ordinary youthful sports. The young philosopher, however, was not at all above accepting his friend's escort past the " eeriest" part of the dark

[1] There is a most interesting room at Inchbonny, though of more recent date than the time which we are describing, in which Mr. Veitch carried on his studies for many years before his death in 1838. His telescopes, books, papers, and microscopes are carefully preserved, and the ancestral relic of Ringan Oliver's sword. Mr. William Veitch, his eldest son, resides at Inchbonny, and inherits his father's taste for constructing telescopes. Another son, whom Mr. Veitch lost in boyhood, gave the finest promise of scientific distinction.

road at the " Scaur," till the outlines of the old Abbey
towers could be seen clear against the sky, when the
trees and their shadows became scantier, and the hooting
of the owls less dreary.  Philosophical as were his
pursuits at this period, he does not seem to have had
entirely an old head upon young shoulders, and many
amusements of his age were participated in.  It is told
of him, indeed, that in the playground as well as in
the schoolroom he took the position of leader,—that
he was ready to face any foe, and that he rarely confessed himself to be vanquished.  For one favourite
sport the means were borrowed from his beloved Inchbonny.  " The ' auld wood,' as it was called, was a grand
scene of boyish exploits in spring.  It was a magnificent wood of Scotch firs, about three hundred years
old, on the estate of Stewartfield (now Hartrigge, the
property of Lord Campbell).  Many of the stems rose
forty or fifty feet without a branch.  These the boys
ascended, with ' speilers'[1] or iron cramps on their ankles,
to reach the crow-nests.  Those who could command a
gun availed themselves of that weapon.  Armed with one
belonging to his friend, Brewster was one day very
valiant in his onslaught of the young French ; when, in
the heat and confusion of the fray, the shot in passing
struck a young Hilson, whose scream ended David's
sport.  He returned to Inchbonny with the gun, crest-
fallen, staying till night was setting in, when it transpired that he was afraid to go home without some
intercession.  The request was complied with, and the
tragic affair became a good joke, no great harm having
been done."

There were other influences besides these scientific

_____
[1] From "speil," to climb, an old Scotch word.

ones, although we know so little of their precise nature that they can only be indicated. The orthodoxy of a very early first love can be traced even at this period. Margaret Somerville, a child of the manse, was its object. The boy-lover often stole from his games in the Abbey aisle to clamber up to an ancient window from whence he could watch her in the manse garden. It seems to have been a deep, tender, and reverential feeling on his side, lasting into his early manhood, although it is supposed that " he made no sign," and that the young lady was unconscious of the impression she had made. She died unmarried in 1843, and was buried near the very window whence she was gazed upon by her youthful adorer.

On the opposite side of the river to Brewster's house there lived a weaver named Robert Waugh, a man of intelligence and humour, and also an excellent singer. The boy, athirst for knowledge, spent many an hour amongst the looms listening to songs, old stories, recitations from favourite authors, and literary dissertations. On one occasion James Scott and David Brewster had each written a poetical effusion, and each believed his own undoubtedly the superior. They agreed to make Waugh the umpire ; having accepted the office, David impressed on him the necessity of reading both compositions aloud with due care and emphasis. The weaver, being fond of a joke, proceeded to recite mournfully an absurd rhyme, to the horror and indignation of Brewster, who had set himself to listen with his usual intent earnestness. Scott enjoyed the joke, and laughed heartily, which made matters worse, but Waugh went up to the angry boy, clapped him on the shoulder, and said, " Never mind, Daavid, ma man, your heid 'll be a frontispiece to a buke when that lad's forgotten!"

A soldier was billeted upon one occasion on the Brewster family, and to this man the future President of the Peace Congress formed the strongest attachment. The depth of the feeling may be conjectured from the fact that, in 1860, when pointing out to his daughter-in-law, Mrs. Macpherson, the grave of Margaret Somerville, he could not visit the soldier's grave with any one —he went to *it* alone. Other elements of adventure also there were in his circle of acquaintance. Colonel Rutherfurd, the laird of Edgerston, and his cousin, " the Major," were men of great intelligence. The life of the latter had been one of most varied adventure, having been for many years in the wilds of America, where he was detained as a prisoner for some time by a tribe of Indians. He lived at Lintalee Cottage, a farm a little above Inchbonny, and it is said that David greatly delighted in his conversation,—he kept up his intercourse with the Edgerston family for many years, visiting there whenever he returned to his native town.

Just opposite the Brewsters' house lived a family of the name of Robertson—the father being the Antiburgher minister of Jedburgh. The intimacy between the families was great, and there was a system of telegraphs carried on between the youngsters of the two houses. A son of Mr. Robertson was afterwards an eminent London solicitor, of the firm of Spottiswood and Robertson. Miss Robertson, the last of the family, and a contemporary of David's, died a short time after her early friend's death, but had survived her faculties for some time, so that no early information could be gleaned from her.

C

At the back of the rector's house there was a large open space, half garden and half orchard, containing many old gnarled pear-trees, and a few stones of a ruin, which probably indicated the remains of a monastic building. At the opposite end was Queen Mary's house, which was occupied in Brewster's youth by Dr. Lindsay and his family, the "sweet Isabella Lindsay" of Burns being one of his daughters. Dr. Lindsay and Mr. Brewster were intimate friends, as well as their families. This house was till her recent death, in the ninety-third year of her age, occupied by a charming old lady, Miss Armstrong, who was in the perfect possession of every faculty except that of hearing. She was a descendant of Riccaltoun, minister of Hobkirk, whose works are well known, and who befriended Thomson the poet in his youth. This lady did not reside in Jedburgh when David was a boy, but her friend Miss Robertson had often told her old stories of him and his family, and some of those early incidents which I have related are upon her authority. Her descriptions of Miss Lindsay (her brother's wife) and Margaret Somerville were very graphic, and given with much animation. Burns's beauty, she said, possessed "auburn hair and violet eyes," was very fair, not very tall, but "*verra sweet.*" Margaret Somerville had the same auburn hair, a small long face, and extreme fairness.

It is interesting to trace in these various intimacies and acquaintanceships the source of much of the social character and vivid variety of interests and sympathies, as well as the fostering of the inventive and scientific genius, which alike distinguished in after life the subject of these " Notes."

# CHAPTER IV.

## THE STUDENT.

WE shall go forth together. There will come
Alike the day of trial unto all,
And the rude world will buffet us alike.
But when the silence and the calm come on,
And the high seal of character is set,
We shall not all be similar. The flow
Of life-time is a graduated scale ;
And deeper than the vanities of power,
Or the vain pomp of glory, there is set
A standard measuring our worth for heaven.
For in the temper of the invisible mind,
The god-like and undying intellect,
There are distinctions that will live in heaven,
When Time is a forgotten circumstance !
What is its earthly victory? Press on !
For it hath tempted angels. Yet press on !
For it shall make you mighty among men ;
Press on ! for it is god-like to unloose
The spirit, and forget yourself in thought,
Bending a pinion for the deeper sky,
And in the very fetters of your flesh,
Mating with the pure essences of heaven !
Press on !—" for in the grave there is no work
And no device." Press on ! while yet ye may !

*From a poem delivered at Yale College,
in 1827, by* N. P. WILLIS.

THE happy days at Inchbonny could not last for
ever. In 1793, at the early age of twelve, David
Brewster went to the University of Edinburgh, and from
that time the visits to Inchbonny became the favourite
holiday pleasure, instead of the daily occupation. In
those days long journeys on foot were undertaken
with an alacrity of which little is known in our times

of easy and cheap transit, and students greatly preferred their own instruments of locomotion to the Jedburgh "fly," which took a whole day to lumber along
between the Scotch metropolis and the Border county-
town.   David used to allude to the walks from Edinburgh to Jedburgh as very pleasant tasks.   After his
forty-five miles' walk home, he generally started again
for Inchbonny the same day, so eager was he to hear what
his friend had been doing, and to recount his own
advances in knowledge.   The separation was further
soothed by a constant correspondence.   Unfortunately
the earlier letters have not been preserved, but those
that remain show it to have been a remarkable one,
and Brewster's handwriting is precisely the same as
that of his later years.   They contain astronomical
calculations, abstracts of abstruse mathematical and
scientific works, notices of the ardent commencement of
his life-long study of optics, as well as of the favourite
amusement of making and testing telescopes and
other philosophical instruments.   There is also frequent mention of the eminent men with whom he was
associated, at first in the relation of student and professor, but marvellously soon in that of friend and companion,—Professor Playfair, Professor Robison, Dugald
Stewart, and others.   Everything, in short, is touched
upon that could interest Veitch and instruct himself;
while the answers he received were well calculated
to keep alive and nourish the desire for scientific and
practical knowledge which had commenced so early.
These extracts give an idea of his experiments and
occupations.   They are taken from the lighter letters
of a series of seventy-two, many of them abstrusely
scientific :—

"EDINR., *November* 15, 1799.

"You may write me this week, and give me any directions which you think necessary for constructing the electrical machine. Would the hair of one's head answer as well for the cushion as horse-hair, as it could be easily got? You will recollect that I once told you that Mr. Edwards used no more than two tools, viz., a Penton-block and a bed of hones. . . . The following composition for reflecting specula, viz., copper 32 ounces, tin 13 ounces, and regulus of antimony $\frac{1}{40}$ of the whole, viz., one ounce to $2\frac{1}{2}$ pound of metal, Mr. Edwards says makes a beautiful metal, *very* like the silver composition, but not quite so white. This composition is infinitely cheaper than the silver one, and might be easily put in practice. . . ."

"EDINR., *Decr.* 6, 1799.

" I have finished the electrical machine, but I cannot make it give a shock. When I cover the rubber with silk, and darken the room, a faint light appears between the rubber and the glass, but when I take away the silk no light at all appears. Might we not infer from this that more electricity would be produced were the rubber to be covered with silk? In this situation it attracts small pieces of paper."

"EDINBURGH, *April* 4, 1800.

" I lately made a very large map of the stars that were near to this planet, which I would have sent you had it not been lost at the Observatory. The man in the Observatory has not yet learned by experience, as we have, what a difficult thing it is to give a speculum a good figure. He is working too for the telescope which I mentioned, but from his careless and unsteady

manner, I can prophesy that none of them will show.
He, however, thinks otherwise. . . . James Scott delivered his lecture in the Divinity Hall here on Wednesday last.  The Professor found fault with it in every
place, and gave it no praise at all.  It was reckoned
the poorest discourse that had been delivered here this
winter.  They even found great fault with his delivery,
in which he so much prides himself.  Sermons that
please the old wives of Jedburgh do not answer when
preached before judges."

"EDINBURGH, *October* 2, 1800.

"I arrived here yesterday at three in the afternoon,
and have been employed since that time in preparing
my Newtonian telescope for seeing the eclipse.  During
the whole of this day the sky has been completely
overcast; it has just this moment brightened up, and
is entirely free of clouds, so that I expect still to see
the eclipse, as it will happen in the space of three hours
from this.  My telescope is in a better state just now
than I ever saw it; and, without the least exaggeration,
shows as distinct as any of the kind I ever saw, though
its magnifying power is 62, and though its aperture
(which is not in the least degree confined) is about
seven-tenths larger than the table prescribes, which
you know makes a great deal of surface.  The calculation of this eclipse, which I made from Ferguson, does
not differ so much from the Nautical Almanac as one
would have expected. . . . The penumbra has just
this moment left the moon; the eclipse began at 45
minutes past 8, and ended 35 minutes past 10, mean
time; consequently the duration has been 1 hour and
50 minutes, the very same as was found in the projection."

An amusing and exceedingly natural feeling of complacency regarding Inchbonny teaching creeps out occasionally, mingled with a little jealousy of superior materials, as in this postscript to the last letter :—

"*P.S.*—I called at the Observatory this evening a little before the eclipse happened, and saw the moon through the two feet and a half Newtonian reflector which the man has sold for ten guineas ; it shows as ill when compared with mine (which is of the same focal length), or with any of yours, as a dirty common refractor does, when compared with a fine achromatic telescope ; nay, to tell the truth, the moon appears far better without it.  It is fitted up in a fine brass tube, and mounted on an excellent stand, whereas ours bear a greater resemblance to *coffins* or *waterspouts* than anything else.                    D. B."

In the following letter there is a touching tribute to his obligation to James Veitch :—

"EDINBURGH, 26, 1801.

" I am sorry you should ever think of attributing my long silence to my ' not thinking it worth while to put off my time in writing to you.'  Far be it from me to entertain such a slight opinion of any of my friends, but particularly of you, whom I have every reason in the world to remember as long as those studies can afford me any delight which you first encouraged me to pursue, and in the prosecution of which I have repeatedly received your assistance.  However, if my memory does not fail me, I believe that I wrote you last, and that I have never received an answer.  Have you heard of the Galvanic Column, which gives a shock in the same manner as the electrifying machine? It is made in the following manner:—Take any number of plates of copper (penny-pieces answer very well), or, which is better, of silver (such as half-crowns or crowns), and an equal number of tin, or, which is much better, of zinc, and a like number of disks or pieces of card or leather, or cloth soaked well in water. Then build up a pile of these, viz., a piece of copper, a piece of tin, and a piece of wet card; then another piece of copper, a piece of tin, and a piece of wet card, and so on till you have a column as high as you please. If you use silver and zinc instead of copper and tin, you must put the silver in the place of the copper, and the zinc in the place of the tin. The instrument being thus completed, you will receive a shock by wetting your hands and applying one of them to the lower plate and the other to the upper one, and that as often as you please to lift up your fingers and put them down. When the hands are not wet, the galvanic influence cannot pierce the dry skin; and in order to receive a smarter shock, it is proper to wet the hands and grasp with them two large pieces of metal, and then touch the upper and under plates of the column with these pieces of iron. Twenty pieces will give a shock in the arms if the above precautions be attended to, and one hundred may be felt to the shoulders. It is also very remarkable, that if you place a piece of zinc above your tongue, and a piece of silver below it, and then bring them in contact over the tip of your tongue, you will feel a curious acrid sort of taste, and if you let the zinc remain on your tongue, and place the silver on your eye so that it may touch the cornea, and then bring their extremities in contact, you will feel a curious taste in your tongue, and see a flash of fire dart from your eye.—I am, yours affectionately,

"DAVID BREWSTER."

In this correspondence there are many hints denoting the profession for which the young scholar was preparing with deep and reverent attention, even in the midst of the charms of science. Mr. Brewster was a strenuous supporter of the Established Church, and destined his four able sons to enter its pale as ministers. James, George, and Patrick each followed this path. The first reached the truest eminence which it admits of. Although the minister of a quiet country parish, he was called by Dr. Chalmers, "one of the four pillars of the Church of Scotland," from his holy life and wise judgment. Like Daniel, he was a man "greatly beloved," and was honoured of God in the conversion of many souls. As an author he was well known, and successful. He was the friend and father, as well as pastor, of the fishing population of Ferryden, a coast village in his parish; and many trace the good work which still goes on there to the example and prayers of their godly minister, who "though dead, yet speaketh."[1] Dr. Brewster was a Disruption minister, leaving without hesitation one of the loveliest churches and manses in Scotland. The second, Dr. George Brewster of Scoonie, was possessed of literary talents, and was one of the *Encyclopædia* writers. He had also considerable musical talent, but he turned his attention principally to parish business, and while preaching excellent sermons, and commanding universal re-

[1] "It's Dr. Brewster's prayers," is a frequent exclamation amongst the old people, when another and yet another conversion to new life takes place amongst the fisher families and their neighbours.

spect in his Fife parish, he did not fulfil further the
promise of his youth. Patrick, the youngest, who
became one of the ministers of the Abbey Church of
Paisley, was gifted with most versatile genius ; but
had that touch of eccentricity which so frequently
mars the finest gifts, and while using them he did not
always refrain from abusing them. Balked in his
strong desire for a military or naval life, and meeting
with one of those disappointments which, unlike the
gentle love-story of David's boyhood, scathe and blight
a sensitive life, Patrick found more congenial excitement in politics than in the calm routine of parish life.
His peculiar gifts were music, painting, and oratory.
Printed sermons and addresses of the finest composition still remain, but he was accused, and not always
unjustly, of turning his eloquence to bad account,
and he was much reprobated for his Chartist and
Radical views. With all his mistakes in judgment,
however, he always used his vast influence over the
people on the side of order and peace, and from this
cause, at last honourably lost it. He was emphatically
a man before his time, and most of the subjects of his
greatest agitation are now received without a question,
such as the repeal of the Corn Laws and the lowering of
the franchise. His riper views, indeed, were not more
Radical than those brought into practice by the Conservative Government of 1866. His abhorrence of Popery
and of every concession made to it was extreme, and
his unceasing efforts to expose the abuses of parochial
power were useful in calling the attention of the public
to the many evils connected with the Scotch Poor Laws.

With the knowledge of his father's wishes, his
brothers' compliance with them, and an evident interest

in theological studies, no different course seems to
have entered into David Brewster's mind; and when
fears of the incompatibility of the two pursuits did
intrude, it was evident that Science was the one he
intended should give way, as we see in the following
extracts from letters to Mr. Veitch :—

"EDINBURGH, *December* 26, 1800.

" . . . The truth is, I have been so much engaged
in studying Divinity since I saw you last, and on that
account have had so little leisure for paying attention
to astronomy and optics, that I have scarcely seen any
new book upon these subjects since I left Jedburgh.
I am happy to hear that you have such a good speculum for your seven-feet reflector, and that it shows so
distinctly. Indeed, I am perfectly convinced of the
truth of your remark, that the best telescopes are produced by frequent trials, and by choosing the best and
rejecting the worst. But though this is the method
which Herschel generally takes, yet I do not think it
redounds greatly to his honour when he tells us that
he often made a hundred without hitting upon a good
one."

"EDINBURGH, *March* 12, 1802.

" . . . I have been so much engaged of late in writing discourses for the Divinity Hall, that it has been
out of my power to answer your former letters, and till
the end of next month I am afraid I shall be in the
same situation. I have never seen the instrument
which you mentioned in your last. There is one of a
similar nature in the Natural Philosophy class here.
When the picture of a woman extremely distorted and
deformed is placed before it, the image in the mirror is

perfect and well-shaped. It is nothing more than a cylinder of the common metal used for reflecting telescopes, polished on the outside. It is only a picture of a certain form which answers, and I believe it is made in the following manner :—The picture of a handsome woman is placed before the cylindrical mirror. The image of the picture will consequently appear behind the mirror very much distorted and out of shape, and if this image is drawn upon paper and placed before the polished cylinder, it will appear handsome and perfect. I wish you would give me a description of the other in your next letter."

"EDINBURGH, *February* 17, 1803.

" My attention of late has been completely turned from my old and my favourite studies, so much so, indeed, that I am sometimes afraid that I shall never be able to fix it upon them again."

In the midst of all the varied study it is pleasant to see how David could turn his whole attention to answering, to the best of his ability, " without assistance," as he himself says, a question put by Mr. Veitch :—

" EDINBURGH, *October* 16, 1801.

"DEAR SIR,—I received yours, and shall endeavour, as far as I am able, to explain the passage of Matthew which you have mentioned. It has long been a dispute among divines whether or not all prophecies have double senses, that is, refer to two events at the same time, and a great many learned men have defended each side of the question. Now, if we believe in the double senses of prophecy, it is easy to explain the whole 24th chapter of Matthew, by saying that it refers, in the first part, to the destruction of Jerusalem, and in the last, to the

end of the world; but still a difficulty occurs in the
34th verse, where it is said that this generation shall
not pass away till all these things be fulfilled. I do
not agree, however, with those who believe in the double
sense of prophecy, as it is contrary to that simplicity
which ought to be expected in the sacred writings, and
would therefore explain the chapter in a different manner, as referring wholly to the destruction of Jerusalem.
In the 3d verse, the coming of Christ and the end of
the *world* (or rather, the end of the *age*, as it is in the
original), signifies nothing more than that period when
the Jewish Polity and State should be completely over-
turned, and the Christian dispensation become more
firmly established, by the destruction of its enemies,
and by the interposition of Christ in the overthrow of
Jerusalem. The prophecy in the 29th, 30th, and 31st
verses, which contain the greatest difficulty, appear to
me to have been fulfilled at the destruction of Jerusalem. Christ might with sufficient propriety be said to
come in the clouds of heaven, in power and glory, when
at that time the most wonderful appearances in the
heavens took place that were ever seen. The stars may
with propriety be said to fall from heaven, when light-
nings and great globes of fire destroyed the workmen
appointed by the Emperor Julian to rebuild Jerusalem;
and, if I am not mistaken, the sun and moon were
actually darkened at the overthrow of that city. All
the tribes of the earth might be said to mourn when
so many thousands of Jews were slain in such a cruel
manner, and when they heard of the dreadful barbarities
which were committed upon them by the Romans. And
the elect might properly be said to be gathered together
from all quarters of the heavens, when the Christian

religion, as it then actually did, extended itself rapidly over most countries of the known world, and brought the glad tidings of salvation to men of every description, nation, and language. This is the only consistent explanation of the passage which I can give without any assistance.—I am, yours sincerely,

　　　　　　　　　　"DAVID BREWSTER."

David's College career was marked by brilliancy as well as solidity. At the age of nineteen he added M.A. to his name, and in the same year (1800) he made his first discovery, an important event in the life of so young a philosopher, which was to be so frequently honoured by the same useful way-marks. It was in his favourite science of Optics, discovering a new and important fact, while submitting the Newtonian theory of Light to a serious experimental test. At the age of twenty we find also the commencement of his independent literary career. Before that period he had regularly written for the *Edinburgh Magazine,* a periodical combining science and literature, but he then became its editor. The volumes that remain contain much interesting matter. The following letter to Mr. Veitch refers to this busy period. Mr. James Fair was one of the young Jedburgh and Inchbonny *literati :*—

　　　　　　　　　" EDINBURGH, *August* 6, 1802.

" MY DEAR SIR,—As I have been very much engaged since the receipt of your letter, in conducting the *Edinburgh Magazine,* and in preparing for going to the country, it has been out of my power to write you sooner. . . . I am very sorry to hear that Mr. James Fair is ill pleased at me for the rejection of his paper. You may render my compliments to him, and tell him

that at that time I had no power either to admit or
reject any paper that was sent to the *Edinburgh Magazine*. His other paper also, on Humanity and Gratitude
to Animals, arrived just before the editorship of the
Magazine devolved upon me, so that for its rejection I
am not answerable. So far was I from wishing to discourage Mr. Fair, that I would have exerted all my
influence with Dr. Anderson to get his papers admitted,
had I not been conscious that the very admission of
them was injurious to the author himself. From his
last paper I observe that he is improving rapidly in
composition, and I am sure he would have been sorry
about a year after this if he saw two of his papers laid
before the public which he knew to be replete with
inaccuracies. Mr. Fair composes as well as any young
man can be expected to do who has not dedicated to
it a great deal of his attention ; and I am sure that if
my first compositions, or those of any other young man
who is not much accustomed to write, were given to the
world, they would cut a very awkward figure indeed.
If you look at last *Edinburgh Magazine,* which you will
get from my father, you will see some interesting scientific notices which I collected from a variety of quarters,
and a curious paper on Ventriloquism, which I inserted,
as it is the first explanation of that curious phenomenon
which was ever given. I set off to-morrow for the
North, and will probably see you at Jedburgh in the
space of three weeks. Excuse this scrawl, as I have all
the literary and scientific notices for next Magazine to
write this forenoon, and believe me your affectionate

" DAVID BREWSTER."

To all these arduous tasks Brewster added those of

tuition. In 1799 he became tutor in the family of Captain Horsbrugh of Pirn, in Peeblesshire, remaining with them, more or less, till 1804. At first he resided with the family in George's Square, Edinburgh, in winter, and at Pirn in summer, but when his eldest pupil went to an English school, Mr. Brewster found it more convenient for his multifarious studies to live in a lodging, and go in the evening to George's Square. One of his pupils, six years younger than himself, still survives, and remembers beginning her letters under his superintendence. He was a great favourite with the children, especially with those who could enter into his pursuits, often amusing and interesting them with his experiments, and especially with his own early favourite, the electrifying-machine. Miss Horsbrugh remembers the starts and the shocks she received, and also being occasionally left in the dark, when Mr. Brewster appeared among them with his outstretched hand and fingers all in an apparent blaze from phosphorus. She also remembers how characteristically busy he was even then with bits of coloured glass, making experiments upon Light, and how he was often late at night observing the stars. Some of his scientific practices greatly incensed Mrs. Dickson, the housekeeper, who declared that he would never rest till he had set the house on fire.

One is thankful to find, even during his student days, some traces of the relaxation and exercise which must have been so needful in such a busy life. In 1802, David made a tour of visits in Fife and Forfarshire, visiting, amongst others, his uncle, Dr. Key, at Forfar, and taking a trip to St. Andrews,—of which old city an interesting account had appeared in the *Edinburgh Magazine*,—little dreaming of the twenty-three years of his later life which were to be spent in its cloistered shades.

At that time the threatened invasion of England by the French was a cause of general fear and excitement. "Boney" was a universal object of detestation, and a bugbear even in the nursery, while the slumbers of adults were broken by uneasy dreams of kindling beacons and secret landings. The following letter, written in 1868, from the Rev. Mr. Ramsay of Tranent, some years Brewster's senior, gives an account of his participation in preparations against this dreaded event :—

"One day at the end of last century, when in the College Court (Edinburgh), I met Mr. Brewster coming up from the Divinity Hall, and said to him, 'There is a public meeting to be held to-day in the Archers' Hall; let us go and see what is going on.' When we came there we found the hall crowded, and a committee was just formed for the purpose of establishing a volunteer corps in connexion with the University, for the defence of the country against the threatened invasion of the first Bonaparte. I stated to the meeting that the committee, however respectable, did not consist of persons then acquainted with the state of the College. The late Lord Brougham, who was in the chair, said that there was a good deal in what the gentleman had stated to the meeting, and proposed that I should be added to the committee, and also requested me, with the consent of the meeting, to mention another to act along with me as a member of committee. I immediately mentioned Sir David, who was standing by me, and after that he had frequent opportunities of waiting upon and consulting persons of influence in Edinburgh and the neighbourhood."

D

Brewster also endeavoured, with more good-will than ability, to prepare for practical personal efforts in the volunteering line. During his holiday visits to Jedburgh it is recorded that he practised shooting with the young volunteers on the Ana, a piece of waste ground close to the Jed, generally devoted to such pursuits, the mark being the opposite scaur, called the Ana also, from its vicinity to the shooting ground. Brewster's reputation as a shot had not much improved since the days of the " auld wood," and he frequently missed the mark. His philosophy however generally furnished him with some apology—a favourite one being that the attraction of the water had deflected the bullet from its proper course. This always produced a burst of amusement and argument, in which on one occasion a determination was expressed to refer the vexed question to James Veitch, their great scientific referee. " No, no," said Brewster with *naïveté*, " don't do that ; he'll be sure to give it against me! "

Verse-making too mingled its lighter thread amongst the sterner warp and weft of the student's daily occupation. The keenest appreciation of poetical beauty, and a wish to clothe his thoughts in that garb, accompanied him through life, but the two following poems are, as far as I know, the only ones which were ever printed. The first is taken from a little volume of songs printed in Jedburgh about forty years ago. Mr. Veitch, although a kind patron of local poets, who generally brought him their compositions, highly disapproved of his young philosophic pupil wasting his time in verse, and used to cast a doubt on the perfect originality of the " Banks of Jed," but this is believed to be a mistake.

" When evening decked in grey appears
  To wrap creation in her shade,
I wander 'mong the fragrant briers
  Down by the silvan banks of Jed.
Memory recalls those joyous scenes,
  Those joyous scenes for ever fled,
When Anna listened to my strains,
  While wandering on the banks of Jed.

The milk-white thorn[1] in yonder vale,
  Still blushing with the twilight red,
Minds me of Anna's glowing smile
  When roaming on the banks of Jed.
When with a sweet and murmuring noise
  The stream rolls o'er its pebbled bed,
Methinks I hear my Anna's voice
  Re-echoing from the banks of Jed.

The breathing gales, which health renew,
  And sweetest fragrance round thee shed,
Recall the heavy sighs we drew
  When parting on the banks of Jed.
But, ah ! no stream nor milk-white thorn,
  No breathing gale can make me glad,
Till Anna to these scenes return,
  To meet me on the banks of Jed."

By this time the influence of the early love had
waned, and another had appeared on the scene, celebrated in the above and in various other poems as
" Anna," and in one as the " Rosebud of Glenae." Mr.
Brewster had been recommended to the family of Pirn
by Mr. Scott, the minister of Innerleithen, who was
married to Miss Chisholme of Chisholme. One of their
two lovely daughters was the " Anna" of Mr. Brewster's
admiration, although her name was the less poetical
one of Agnes. She died at an early age, and David
was much affected. He wrote a poem of twenty-four
stanzas on the occasion, which was published in the
*Edinburgh Magazine,* and also printed separately ; the
only copy of it in the latter form which is known to

[1] An old thorn-tree, now rather stunted, in the field beside the Jed, opposite the scaur, near Bonjedward.

remain is retained in the family as an interesting relic
of the past. It was kindly lent to me by Miss Scott
Chisholme, who also showed me the miniature of her
beautiful aunt. The following verses are extracted
from it; and it is touching to observe that the everyday Scotch appellation is used instead of that of the
less earnest lines,—Nancy being the Scotch "diminu-
tive" for Agnes as well as for Anne :—

> Soon shall verdant Spring, returning,
> Waken flowerets from their tomb ;
> Soon shall *Sol*, with fresh flames burning,
> Brighten up the deepened gloom.
>
> But, alas ! no Spring, returning,
> Nancy's eyes can e'er relume ;
> But, alas ! no ruddy morning
> Can recall her orient bloom.
>
> On her pillow flowers may flourish,
> On her grave the sun may shine ;
> But can these our spirits nourish ?
> Can they wake her form divine ?
>
> As for me, I'll ceaseless wander
> Round her verdant, hallowed grave,
> Where *Leithen's* crystal streams meander,
> Joining *Tweed's* proud classic wave.
>
> Ere the sun, from yonder mountain,
> Tip these hills with orient hue,
> I'll repair to sorrow's fountain,
> There to bathe my eyes anew.
>
> Hope's bright eye, in time of trouble,
> Views with triumph yonder coast,
> Where our bark, forlorn and feeble,
> Shall no more in storms be tossed ;
>
> Where Disease, and Death, and Anguish
> Shall be banished our abode ;
> Where our joys shall never languish,
> With our friends and with our God.
>
> Cease, then, mourners, cease complaining,
> Cease deploring Nancy's lot :
> On that coast she now is reigning,
> Evil banished, pain forgot.

# CHAPTER V.

## SETTLING IN LIFE.

Thus Genius, mounting on his bright career,
Through the wide regions of the mental sphere,
And proudly waving, in his gifted hand,
O'er Fancy's worlds, Invention's plastic wand ;
Fearless and firm, with lightning-eye surveys
The clearest heaven of intellectual rays ;
Yet in his course though loftiest hopes attend,
And kindling raptures aid him to ascend
(While in his mind, with high-born grandeur fraught,
Dilate the noblest energies of thought ) ;
Still, from the bliss, ethereal and refined,
Which crowns the soarings of triumphant mind,
At length he flies, to that serene retreat,
Where calm and pure the mild affections meet,
Embosomed there, to feel and to impart
The softer pleasures of the social heart.

MRS. HEMANS.

UP to this time, as we have already seen, and for five years subsequent, no doubt occurred to Brewster's mind of what his vocation was to be,—the Church his field of work, and Divinity his principal study. He had a sincere attachment to the principles and constitution of the Established Church of Scotland, and a thorough acceptance of her doctrinal standards. He had a clear sight to her deficiencies of discipline, however, which in those days were obvious to every observer, and often recalled with horror, in after years, the Socinian doc-trines which in the south of Scotland, were openly held by ministers high in popular esteem,—while other and more glaring derelictions were appalling to one whose own character was described at this time by a young

Roxburghshire *roué*, who had watched his student career, "as the only virtuous character he had yet met among young men."

Brewster had evidently written complaints of Dumfriesshire upon this topic to his friend of many years, Dr. Andrew Thomson, then minister of Sprouston, who replied by giving painful evidence that the clergy in his district were no better than in Annandale, concluding with this sentence :—" Your clergymen and bailies must be very queer. We have some in this country that are not far behind your Annan parsons in attachment to the bottle. Though I have not yet heard of any getting a *tankard* into the pulpit, yet we have had some instances of its *contents* being carried there in high style. Indeed, this is a singular Synod, and, with reverence be it spoken, is more celebrated for anything, even drunkenness, than for divinity. My friend Mr. G— of L— says that it is the paradise of the Church; and why? because here a clergyman may do whatever he pleases !" All this did not discourage Brewster from entering the Church, but it inclined him to that better side, although a minority, which then received the name of the " Evangelicals," or " wild men," as their Moderate friends termed them. Of this party Dr. Andrew Thomson was long the avowed leader; and although in many cases it was but a name, and politics mixed largely with its profession, yet it certainly comprised most of the godly ministers of the Scotch Church. Dr. Thomson says in another letter to his friend :— " Nothing would give me greater pleasure than to see you settled in this neighbourhood. We stand in need of ' wild men' and men of respectability. The Church is wofully deficient in this quarter of the country."

The erroneous impression that Brewster possessed no pulpit gifts, only preached once, and on that occasion failed, has become very general. The following account will be read with interest. It is taken from the recollections of the Rev. Dr. Paul, now one of the ministers of the West Kirk, who was present as a boy, and on whose mind the whole scene made a deep impression, being a somewhat singular one for a "first sermon :"—

"Having passed through the Hall, and having been licensed by the Presbytery of Edinburgh, Brewster preached his first sermon in St. Cuthbert's, or, as it is familiarly termed, the West Kirk of Edinburgh, one of the largest in Scotland, accommodating 2500 hearers. This spacious church, with its double tier of galleries, was on this occasion unusually crowded, for the reputation he had acquired drew together, not only numbers of his fellow-students, but of literary and scientific men, anxious to hear how he would begin his professional career. The ministers were present, Sir Henry Moncreiff and Dr. Paul,[1] and the appearance of that vast congregation, which the youthful preacher was to address, was most imposing, and, to a person of his anxious, nervous temperament, must have been most formidable. He ascended the pulpit, and went through the whole service, for a beginner evidently under excitement, most admirably. He had his discourse thoroughly committed to memory, and delivered it with great energy, increasing to the close, which was in these words, 'Let it be our firm resolution, our constant endeavour, our importunate prayer, that so long as we have being and breath, we will serve the Lord.'"

[1] Father of the present Dr. Paul.

The following extract gives his own brief and modest
account of this crowded church and favourable commencement. Writing to Mr. Veitch, March 16, 1804, he
says :—" I wish you would give me a description and
figure of your instrument for finding the stars in the
day-time, and I will take notice of it in my paper on
the Progress of the Arts and Sciences. I am at present writing notes and an appendix to a new edition of
Ferguson's Lectures. If any remarks have occurred to
you when reading the mechanical part, I wish you
would mention them in your next. As I am now
licensed, I have not so much time to spare for philosophical studies as formerly, and anything new which
I meet with is generally in my paper in the *Edinburgh
Magazine*. I preached my first sermon on Sunday last,
in the West Kirk here."

After this period he preached frequently in Edinburgh, Leith, and elsewhere; and his ministrations,
judging from various independent testimonies, seem
always to have been most acceptable from the beauty
and earnestness of his style, and his well-known gift of
creating interest out of the driest subjects. But if
pleasant and profitable to others, these ministrations
became more and more the source of great pain and
discomfort to himself. Even before his first appearance
at St. Cuthbert's, he must have suspected his own nervous infirmity, for he is recorded to have expressed himself " as resolved to try the worst at first," and he never
forgot the intense, though suppressed, suffering of the
effort. Success did not make matters easier, for he
never preached without severe nervous restraint. A
delicacy of health—owing, there is little doubt, to his
early and constant application to study, although it

quite left him in later years—showed itself at this time
in nervous faintness, only occurring when making any
public appearance. Consciousness of this weakness
and fear of failure made him also somewhat sensitive
as to the opinion of others. The Rev. Mr. Ramsay
tells the following anecdotes :—being together at a
large dinner-party, " when we met in the dining-room
Mr. Brewster was asked to say grace. He began,
but as he went on the words choked in his mouth,
and he sat down in a faint. He was taken out of
the room, and I went along with him. By proper
attention he soon recovered, but we both lost our
dinner with the party ; and I said to him, ' Catch
me, Brewster, if I'll ever dine with you again.' On
another occasion, when he was engaged to preach in
St. Andrew's Church, Edinburgh, I was present, and
met him when the sermon was over. He asked what
the people were saying of him. I told him, by way of
derision, that the people said that they never heard
such a bore in the pulpit; upon which he said that
the people knew nothing about preaching, and declared
that he would never preach there again."

In 1804 Mr. Brewster entered the family of General
Dirom, of Mount Annan, in Dumfriesshire, with whom
he remained as tutor till 1807. Part of each year they
resided at Murrayfield, probably in the old mansion
which now contrasts with the beautiful new villas and
gardens which nestle at the foot of the finely wooded
hill of Corstorphine, in the immediate neighbourhood
of Edinburgh. His work of tuition does not appear to
have interfered with his own studies and literary and
scientific pursuits, and yet that he performed it in no
perfunctory manner may be gathered from the warm

and long-continued esteem with which he was regarded by General Dirom and his family. He continued scientific and general correspondence with the former for many years. While at Murrayfield, ghost-stories got afloat, from the singular fact of a figure in flowing robes being repeatedly seen dealing with invisible and warlock powers,—Brewster, in his dressing-gown, observing the moon and planets, being too commonplace a solution of the mystery to be acceptable to the superstitious !

His inventive genius was not idle at this busy time. He writes to Mr. Veitch :—

"MURRAYFIELD, *Dec.* 9, 1805.

" MY DEAR SIR,—I am so much ashamed of my delay in writing you that I will not attempt to make the least apology. I may assure you, however, that my mind has been so much occupied, and even distracted, by a variety of almost opposite pursuits, that I have corresponded only with my father and brothers for these two years past. You will probably have seen that I noticed your new plough in one of the late numbers of the *Edinburgh Magazine.* . . . From the little knowledge which I have of the construction of a plough, I cannot form an opinion of the merit of your improvement without seeing the models, but I have not the least doubt that ploughs of your construction will greatly excel those of the common form. . . . I have lately invented two new micrometers, the one for terrestrial, the other for celestial purposes. The first consists of a circular piece of mother-of-pearl, placed in the focus of the eye-glass of a telescope, and forming the field-bar. The mother-of-pearl is divided at its circumference into 360 equal parts, each part in the one which I got made being only $\frac{1}{250}$ of an inch. After finding by experiment the angle subtended by the whole breadth of the micrometer, it is easy to compute a table by trigonometry, which will show the value of any number of degrees. This micrometer is vastly superior to Cavallo's : it does not obstruct the field of view ; it does not require to be turned in the focus of the eye-glass ; and it has the advantage of a larger [*word wanting*]. It may be used also for celestial purposes, and particularly for making a map of the moon and measuring the spots of the sun. The other micrometer, for celestial purposes, is more complicated, and would require a figure for its explanation. It might be made for a few shillings by any workman, whereas micrometers of the common kind, though less accurate, cost about thirty guineas. . . .

" By the bye, the last star but one in the tail of the Great Bear, which has the little one beside it, is a fine double star, which any of your glasses will discover. The first star of Aries is also double. I discovered them lately, but found to my mortification that they had been observed long before.—I am, dear Sir, yours sincerely,

" DAVID BREWSTER."

About the same time we find in contemporary letters that he was spoken of as a candidate for the Chair of Mathematics in the University of Edinburgh, vacant by the resignation of Professor Playfair. He received warm good wishes from many zealous friends, amongst others from Sir William (then Dr.) Herschel, with whom he had by this time, as with many other learned men, commenced correspondence,—-Dr. Herschel, from the great similarity of tastes and pursuits, having long been

a great object of interest to the Inchbonny friends.
Brewster's name does not appear in any document as
a candidate, and it is probable that he did not come
forward when he found that so eminent a man as John
Leslie appeared upon the scene. That he should even
have been spoken of for such an important position at
so early an age is a proof of the standing which he had
already acquired.

The year 1805, owing to this academical contest,
was one of great interest and excitement in Edinburgh,
one of those storms of controversy taking place, which
it is hoped do good by clearing the mental atmosphere,
although the noise and the heat at the time seem but
to injure and scathe. Mr. (afterwards Sir John) Leslie
was apparently the most natural candidate, from his
acknowledged position, not only as a mathematician,
but as a practical philosopher and scientific author.
A minister of the Established Church, Dr. Macknight,
son of Dr. Macknight, the well-known harmonist of
the Gospels and commentator on the Epistles, and himself, it is said, an accomplished man and able mathematician, but little known to fame, came forward
to contest the chair. His cause was warmly espoused
by the Moderate party of the Scotish Church, and the
"ministers of Edinburgh" became pitted against the
men of science. According to the ancient constitution
of the University, of which they were chiefly the
founders, they had a right to advise the Town-Council
in the appointment of professors, and this right they
now attempted to revive, two of their number, supported by the rest, formally declaring that they would
use all legal means to prevent the appointment of Mr.
Leslie. Falling back upon the well-known standards of

their Church, and ignoring the notorious breaches of
orthodox doctrine hitherto winked at in Scotch pulpits,
a cry of "Heresy" was raised against Mr. Leslie and
his party. Some incautious statements which Mr. Leslie
had made as to Cause and Effect, in his work entitled
*An Experimental Inquiry concerning the Nature and
Propagation of Heat,* were attacked bitterly by the
newly-constituted censors of doctrine, which were on
the other side explained and defended with philosophical ingenuity. Many of the Evangelical party took
the side of the philosophers, on the ground that pastors
of the Church ought to find sufficient occupation in
the care of their flocks, without striving either for the
honours or the emoluments of secular professorships.
So the war raged, one of the principal weapons on both
sides being, as is usual in such cases, a series of the
most acrimonious pamphlets. During the hottest of
the fray, an anonymous pamphlet appeared, so different from all the others that it took both parties by
surprise; it was entitled *An Examination of the Letter
addressed to Principal Hill in the case of Mr. Leslie, in
a Letter to its Anonymous Author, with Remarks on
Mr. Stewart's Postscript and Mr. Playfair's Pamphlet,*
—By a CALM OBSERVER; and it professed to adopt "a
mode of discussion remote from personal invective."
This pamphlet was by David Brewster; it is one of
his ablest productions, and has indeed been called
"one of the finest pieces of satire ever written." It
professed to take the part of "the ministers of Edinburgh," looking at the subject from their own point
of view, and fighting them with their own weapons
so skilfully, that at first they believed that it was
written by one of their own party. Dr. Andrew

Thomson, himself one of the Lesleian pamphleteers, thus writes to Brewster :—

"I received the parcel containing your letter and the two pamphlets. I was delighted with the *Calm Observer*, and indeed laughed with him for a whole night. Whoever he be, I rejoice that he has thought proper to publish his lucubrations on the Leslie controversy. He does not stick very close to his text, but he wanders with a good grace, and is always the more entertaining the further he is from the subject. I understand (and this circumstance gives me most pleasure of all) that Leslie's opponents have been completely taken in. They were full of expectation, and—oh, glorious change !—are now full of disappointment and mortification. I perceive that the *Calm Observer* has read Junius with advantage. His transition from the humorous to the serious is happier than I should have looked for. The poetical patches are exquisitely appropriate. The irony, especially to a Lesleian, will be irresistible."

It is difficult to extract what would be intelligible from such a work of local satire, but the following short quotation may give some idea of it. Professor Playfair had published a pamphlet in which he took the practical view of the subject held by those who thought that clergymen should have no time for secular callings :—

"Mr. Playfair's reasoning is at best an *argumentum ad hominem*, which can never be extended to the polite metropolis of Scotland. Here there are no sick to visit, no dying to pray for, no children to examine, no ignorant to instruct, no sinners to convert ; and if Mr. Playfair was so old-fashioned as to revive the antiquated usages of our well-meaning forefathers, let him not imagine that gentlemen of liberal education and

polished manners will stoop to the pious drudgery of
a country curate. . . . If the Moderate party in our
Church have hitherto been indifferent to the great interests of religion, if they have neglected the spiritual
concerns of those whom Providence has placed under
their eye,—it is surely high time that their zeal should
begin to burn. If the dying have died without consolation, the ignorant without instruction, and the
guilty without alarm, it is surely time that a sense of
duty should animate their exertions. You will tell us
that the Moderate clergy attack, both from the pulpit
and the press, the doctrines of that Confession which
they have sworn to defend ; and yet you ridicule the
expressions of zeal when conscience has begun to mutiny
and sting. You will tell us that they countenance the
profane indecencies of theatrical exhibitions ; that they
have devoted themselves with more assiduity to the
injunctions of Lord Melville than to the service of their
Saviour ; and that they have polluted the Assembly with
a shoal of young elders, notorious for their infidelity and
dissipation ; and yet, by a strange inconsistency, you
rail at the very first symptoms of dawning reformation.
If the Moderate party have really been callous to religious impressions, let their zeal against Mr. Leslie now
hallow them in our eyes. . . . It may be said, indeed,
by some, that the sudden conversion of individuals is
always liable to suspicion, and others may imagine
that the instantaneous conversion is more suspicious
still, but I would entreat such persons to consider, that
though vice only has been regarded as infectious, there
is a contagion in virtue which often spreads with irresistible rapidity, and regenerates and reforms wherever
its influence extends."

On the 12th of March 1805 the Town-Council unanimously elected Mr. Leslie as Joint-Professor of Mathematics with Dr. Ferguson, who had before been the senior colleague of Professor Playfair. The Council apparently had carried matters with a high hand, for there was no competition, and Dr. Macknight's name is not mentioned in the minutes. An attempt was made, however, to set this election aside, an overture being sent by the indefatigable "ministers of Edinburgh" to have the subject discussed at the General Assembly of that year; but sensible and eloquent speeches by Sir Harry Moncreiff and other members of Assembly turned the tide of clerical opinion. The overture was dismissed, and the Lesleians left in peace.

It is supposed that Brewster regretted his anonymous publication, notwithstanding its flattering reception, and I never heard him refer to it. A gentleman[1] who saw much of him during the very last years of his life, thus writes :—" I am pretty sure that during the visit to which I have referred I spoke to Sir David of the celebrated Leslie controversy, being well aware of the distinguished part he had taken in it. Neither on this occasion, however, nor on any other, did he seem inclined to speak of this matter, but left me under the impression that he did not desire to be recognised as the ' Calm Observer.' Of course I did not attempt such recognition, but I am convinced that he regretted the personalities in the pamphlet, especially those affecting Dr. Brunton, whom he once rather *sought* an opportunity of describing to me in complimentary terms."

---

[1] The Rev. Dr. Phin of Galashiels, a member of the University Court of Edinburgh.

In 1807, Brewster became a candidate for the Chair of Mathematics in St. Andrews, but did not succeed. In the same year scientific honours began to pour in; he was made LL.D. of the University of Aberdeen, and M.A. of Cambridge, and in January of the following year he received an intimation from Professor Playfair that he was elected a non-resident member of the Royal Society of Edinburgh. This was also a memorable time in his literary history. We are told that one day he was walking in Princes Street with the Rev. Mr. Ramsay of Tranent and the Rev. Mr. Stuart of Bolton, when the former, knowing his literary taste, dropped a casual hint how much a good and thorough Encyclopædia was needed. One can imagine how this spark would kindle up the remaining energy of which so much was already in daily, hourly operation. In the same year he commenced this great undertaking, and soon after he wrote to Veitch, to whom, in his happy retreat at Inchbonny, the heart of his early friend seems ever to have turned :—

"EDINBURGH, 9 NORTH ST. DAVID STREET,
*May* 17, 1808.

"DEAR SIR,—You will probably have heard that I have been lately engaged in a new Encyclopædia, of which the two first numbers have just been published. We are now at the article AGRICULTURE, and are about to send the drawings of the ploughs to the engraver. It occurred to me, however, that it might be of use to you to have a drawing and description of your new plough inserted in our work, particularly as the article Agriculture is written by Mr. Brown of Markle, the editor of the *Farmer's Magazine*, and one of the most experienced agriculturists in the kingdom, and his recom-

E

mendation would be very powerful among farmers. If, therefore, you could send me by post, as soon as possible, a drawing and description of your plough, I shall with pleasure publish them in our Encyclopædia. I intend also to publish in the same article a curious paper by Mr. Jefferson, President of the United States, on a plough-ear, which offers the least possible resistance. General Dirom mentioned to me a few days ago that Lord Cathcart was speaking to him about your plough. Both he and the General are anxious to get something done for you by laying the invention before the Highland Society. The drawing which you sent to me, if it is agreeable to you, may also be used for this purpose. I have been lately attending much to optics, and have invented several new instruments, which I hope to have an opportunity of describing to you, as I intend to be in Roxburghshire in the course of the summer. Excuse this scrawl, and believe me, dear sir, most sincerely yours,                    DAVID BREWSTER."

Notwithstanding the magnitude of this enterprise, and the pressure of his other literary and scientific pursuits, Brewster's firm intention still was to keep them subservient to the other and greater work to which his life had been destined, and he laboured patiently and bravely against his severe nervous trial. In 1808 he applied for the living of Sprouston, vacant by the removal of his friend Dr. Thomson to Perth, and it was presented to him in the September or October of that year by Sir James Innes Ker. At this time, however, the Dukedom of Roxburghshire was under litigation, and a competing presentation was made by the Duchess of Roxburgh and General Ker in favour of the Rev.

Ninian Trotter, who ultimately became minister of the parish. Brewster withdrew his claim the following year, on account of the anxiety and expense of asserting his rights, and the pain and discomfort of keeping the parish so long unsupplied. It was not, therefore, till 1809 that he felt himself free to follow the career so manifestly opening before him. Many said, however, that by so doing, he was "blasting his prospects for life."

I have not found a record of his first visit to London, but long afterwards he told Sir James Simpson that he had been in the metropolis in 1802 or 1803, when he had seen much of Cavendish at the Royal Society Club; he mentioned also having dined with that somewhat eccentric philosopher; a rare event, if we may judge from a story my father also told Sir James. Cavendish invariably had a leg of mutton for his solitary dinner; on one occasion he announced to his servant that six gentlemen were to dine with him that day. "What am I to give them for dinner?" ejaculated the factotum in dismay; "one leg of mutton won't do for six gentlemen!" "Then give them six legs of mutton!" was the philosophical reply.

In 1809, however, we have interesting though brief records of a visit to London, and a short tour in England, as well as of his Edinburgh life, in a small volume for that year of *Kearsley's Pocket Ledger*, in which he jotted down a curious medley of household expenses, social engagements, and scientific discoveries. Leaving Edinburgh on January 10th, he made the journey in easy stages, "examining," as was always his wont, everything of interest. Thus we find that at Chester he "dined with Mr. Fletcher, and examined the shot-

manufactory, and that of white and red lead;" that
between Chester and Wrexham he met "a Welsh
hearse, painted blue," which seemed to have made a
lively impression on his mind, though the remarks
are illegible ; that he "examined Wrexham Cathedral," and "the locks of the Ellesmere Canal," and
"invented the reflecting goniometer," followed by an
entry of travelling expenses.  His first visit to Oxford
is dismissed with very brief record, and his London
life left little time for more than entries such as these :
—"Dine at Salopian.   Drury Lane Theatre,—Man and
Wife,—Bluebeard."   "Saw two Russians and Colonel
Waxall."   "Masquerade in the Opera-house."   "Colonel
Waxall and two Russians dine with us."   "Dine with
Lord Selkirk."   "House of Lords.   Dine at home with
Thomas Campbell.   Sup with Archdeacon Corbett."
"Saw Drury Lane burned to the ground."   "Went to
Woolwich.  Captain Pasley dines with us."   "Discovered
demonstration of lever."   "Dined at Blackheath with
Mr. Groombridge and Mr. Troughton."   At the end of
the Diary there are longer abstracts of what he heard
and learned at different places of interest, some being
headed "Memoranda for Scotland."   As a specimen of
these I give one jotting, evidently put down after this
visit to Blackheath :—"Mr. Troughton informs me that
the spider's web is the best substitute for wires.   It is
perfectly round and elastic, and may be easily fixed
with lacquer varnish.   The kind to be used is the
stretcher by which the spider fixes her web in a given
position. This is to be taken off with a pair of stretched
compasses, and wound round their legs.   One of the
filaments is then held above the field-bar, where it is to
be fixed, and by pressing it against the groove where the

lacquer is placed, and holding it there for a short time, the operation is completed." Judging from the number of social invitations, his circle of acquaintance in London, even at this time, must have been very large. He left its attractions, however, on Monday, March 21, arriving at Cambridge the same day " through Epping Forest. In-vented between Woodford and Epping the Katadioptric telescope." The next day he " dined and supped with Dr. Clarke, Mr. Walpole, and Mr. Woodhouse. Thought of the telescope with two semi-lenses of different foci. Wrote paper on Borckhard's telescope for the *Phil. Mag.*" At Manchester, among other engagements, he " supped at Dr. Mitchell's with Dr. Henry; examined case of conical cornea ; examined Dr. Henry's laboratory, and Mr. Lea's house lighted with gas." " Examined Carlisle Cathedral, April 29th," and arrived in Edinburgh the following day. The Edinburgh jottings contain very brief records of engagements at dinner, supper, and " chess " with the literary and scientific circles of Edin-burgh, of which he was now an acknowledged member, with occasional returns of the same festivities " at home." On one of these last occasions he notes— " When in company with Mr. Anderson and Mr. Camp-bell invented Thermometrical Pendulum for mean tem-peratures." On August 18th he again started on a little tour, going to Dumfriesshire to pay some visits, amongst other places to Mount Annan, and with General Dirom he proceeds by Kendal, Penrith, Ambleside, and Manchester to Buxton. From thence " go to Cas-tletown,—visit Peak and Speedwell Mine with General Dirom, General Stuart, and Dr. Hutton," and on another day " the Dean coal-pit." On September 9th, " break-fast at Warrington and visit plate-glass manufactory,"

returned to Edinburgh on the 12th, and on the 18th
" invented another new goniometer."

Two very brief records are entered about this time,
which nevertheless proved the commencement of a new
and happy era in Brewster's life. On September 27th
he writes—" Dine at Mr. Prentice's at Portobello, with
Mr., Mrs., and the Misses Macpherson ; " and on October 3d, " Dine at Mr. Macpherson's, Portobello." Mr.
and Mrs. Macpherson were on their way to a lengthened
tour on the Continent, and the two sisters, whose
home was with their brother in Inverness-shire, were
to spend the winter with Miss Playfair, sister of the
learned and popular Professor Playfair, Brewster's former
Professor, and now intimate associate. Juliet, the
youngest sister—five years younger than David—was
beautiful, and of true and sterling character. From
that time, in the midst of such entries as " Thought of
new theory of the sun," " Invented new method of
measuring crystals by laying a small reflector on their
surface," " Read paper before the Royal Society,"
" Proposed new theory of meteoric stones to the Club,"
we find suspiciously increased and very frequent references to dining, supping, and drinking tea at ".Miss
Playfair's." The diary for 1810, indeed, contains very
few other records except this frequently reiterated one,
till at last the briefly told conclusion comes, " July 31st,
1810—MARRIED. Set off to the Trosachs " with Anne
Macpherson, the bride's-maid, and Dr. Andrew Thomson,
who performed the ceremony,—such social travelling
after weddings being the custom of the day; and in
more formal documents it is recorded that on that day
he married Juliet, youngest daughter of James Macpherson, Esq., M.P., of Belleville, better known as "Ossian

Macpherson." By this marriage, as his wife was very young at the time of her father's death, he was curiously linked, as has been recently remarked, "to the past, now more than a hundred years ago, when Johnson, Hume, Blair, and others, disputed so acrimoniously whether Ossian's poems were true or false, ancient or modern." As there is at present a revival of this long-mooted subject, it may not be out of place to give a brief answer to the often-asked question of what the private belief was of Brewster and the Macphersons. They never had a moment's doubt as to the complete and entire authenticity of the poems. The originals, they were fully persuaded, had been received by Mr. Macpherson in most cases by oral recitation, and in others from MSS. which had been written down two or three centuries before from the old Highland bards, whose predecessors had sung them long before such innovations as pen, ink, and paper were known amongst the Celts. This was earlier, however, than many "Sassenachs," with Dr. Johnson at their head, believed possible. Sir James Foulis, in an interesting letter to Mr. Macpherson, dated "Colinton, May 22, 1784," says, "It cannot be disagreeable to you to learn that there is extant a book in Gaelic, printed in 1567. I presume it is the first book in that language that ever was printed. I have now before me a folio of nearly seven hundred pages translated from Latin into Gaelic, about two hundred years ago. It treats of Medicine, and the transcriber got sixty cows for his pains—a proof that the art of writing was then rare in the Isle of Skye, but not unknown." [1]

Though Mr. Macpherson was a good prose-writer and

[1] Unpublished letter.

historian, his published attempts at versification were
not considered sufficiently successful to give any
promise of originating such noble poetical imagery as
is to be found in the ancient bard, although his genius
might be quite able to do justice to a translation—
which in some instances might be a tolerably free one.
Mr. Macpherson never sought to deny this for a
moment. In a letter to Mr. M'Lagan, he writes :—" I
have met with a number of old manuscripts in my
travels; the poetical part of them I have endeavoured to
secure." Mr. Macpherson was so far from improving
upon Ossian, however, that the very imperfections of his
translation are the strongest proofs of their authenticity;
in an old newspaper of nearly fifty years ago, which I
have before me, I find it stated, " that Macpherson was
not able to give the fire of the original, and fell into
grievous faults." The faults were principally from his
imperfect knowledge of the intricacies of his native
language, having turned his attention at a very early
age principally to classical studies. We are told that
his translation of a passage in Fingal, " spirited so far
as it goes, falls so far short of the original in the picture it exhibits of Cuchullin's horses and car, their har-
ness, trappings, etc., that in none of his translations is
the inequality of Macpherson's genius to that of Ossian
so very conspicuous." [1] One whole passage of one of
the original poems, in which Ossian gives a noble description of Fingal's ships, of nautical feats, and terrific
sea-storms, Macpherson omitted altogether ; which
omission is thus commented upon :[1]—" I can account for
it in no other way than his having been born in Bade-

[1] Quoted in a book entitled *Notes on the Authenticity of Ossian's
Poems*, printed, but not published, in Edinburgh, 1868, by a member of
the Society of Antiquaries of Scotland.

noch, one of the most inland parts of this kingdom, where, not having access to, he was unacquainted with, that kind of imagery ; he did not therefore perhaps understand the original passage."

Although Macpherson had to bear the brunt of accusations of forgery and imposture, he by no means stood alone in his attempts to introduce the beautiful poetry of the North to the ungrateful public of the South. The Dean of Lismore collected and translated, between 1512 and 1529, upwards of 2500 lines of Ossianic poetry, taken principally from recitations. Mr. Jerome Stone, a schoolmaster at Dunkeld, commenced a collection of ancient Gaelic poetry, but died before his collection was finished; he published, however, a translation of poems in 1756. Mr. Duncan Kennedy completed a collection, in 1785, of twenty pieces, giving the names of the reciters. The Rev. Dr. John Smith published, in 1780, the translations of fourteen Gaelic poems collected by him, which it is said are " all remarkable not only for the same strain of high and impassioned poetry, but also the same delicacy and refinement of sentiment, which form so remarkable a feature in the poems translated by Macpherson." It is remarkable how many clergymen interested themselves in this national subject. The Rev. Alexander Irvine of Little Dunkeld, when a Gaelic missionary, made a collection of Ossianic poems from recitation, which are, many of them, the same as those published by Macpherson, with only the variations of having been given by different reciters. The Eastern habit of recitation, which long prevailed in the Highlands, even up to a recent date, was little known or done justice to by those who came

" To fight Macpherson in his native north."

The following is part of an unpublished letter from
Mr. Alexander Small to Mr. Macpherson :—

"No. 120 High Holborn, *Octr.* 11, 1781.

"Dear Sir,—On Shaw's joining in Johnson's continued and obstinate abuse of Ossian, I wrote to a
relative of mine who preaches in the Earse language,
expecting that he would give me some account of Shaw,
and of the authenticity of Ossian's poems. He excuses
himself as not a sufficient judge, but refers to a friend
of his, Mr. Maclagan, minister of Blair-Atholl. Mr.
Maclagan writes as follows :—' That Oisein and Fionn
were often in Ireland no one denies, but that no more
makes them Irishmen than your humble servant. I have
a copy of *Murbhadh Chonnlaich le Cṷthulluim* that I
got in Ireland, and (it) mentions both these men as
strangers from the East. The copy is much inferior to
and different from the Scotch copy got in the parish of
Mullin. That Oisein's poems constituted great part of
the Highlanders' usual entertainment over-night is a
thing well known to every one that knows anything
about the Highlanders at all, and the further back he
goes, it seems the more so. Mr. Macpherson has translated several passages of poems I sent him myself, and
Cargon More almost entire. I have several more of
them. There are still here men who repeat part of
them. Had A. S. come to Scotland last summer, as I
hear he intended, he might have heard part of them
repeated, and seen them written and translated upon
the spot. He may come next summer, and if these
men and I live, he may have the testimony of his eyes
and ears.' "

# CHAPTER VI.

## NOTES OF LIFE FROM 1810 TO 1814.

" IF I do this, what further can I do ? "
"Why, more than ever. Every task thou dost
Brings strength and capability to act.
He who doth climb the difficult mountains
Will the next day outstrip an idler man.
Dip thy young brains in wise men's deep discourse,
In books which, though they freeze thy wit awhile,
Will knit thee in the end with wisdom."
BARRY CORNWALL.

THE choice of life-work which Brewster made was
laborious in the extreme, and he might have said, with
Wesley, that " Leisure and I have taken leave of each
other." Yet it happened with him, as with all active
natures, that the more he did the more he found time to
do. The dying words, long years afterwards, of his friend
M. Arago, might have been taken at this time, as ever
after, for the key-note of Brewster's life, " Travaillez,
travaillez bien." During these four succeeding years
the *Encyclopædia* bulks largest in his daily work, and
on till the publication of the last and eighteenth of its
bulky volumes in 1830, it continued a most anxious
and arduous undertaking. Indeed, its unsatisfactory
pecuniary results, and a long and painful lawsuit arising
out of it, weighed heavily upon his whole future life,
and was not removed till within a few years of its close,
when a compromise took place. The work itself commanded great admiration, and still holds its own recog-

nised place, but the extreme irregularity of publication marred greatly its success at the time. This was owing principally to the dilatory conduct of its literary contributors. It was not only that the discomfort and the unpopularity fell heavily upon the editor—himself a man of punctual habits of work,—but also the unavoidable labour necessary to make any way at all against such negligence and delay. Many of his friends promised articles and forgot all about them. Mr. Stuart of Bolton having engaged to write, hired a room in the High Street to be near books of reference, but even after such demonstrations of diligence no article was forthcoming, and the editor was obliged, as usual, to do the work himself. The mere correspondence of the *Encyclopædia* would have been work enough for most men —many of the letters requiring editorial responses being of a trivial and unnecessary kind. He thus writes to Veitch :—

" JESSEFIELD, PORTOBELLO, *Sept.* 6, 1810.

" I would have answered (your letter) long before this had I not been prevented by a load of business. You will easily understand the nature of my situation, when I mention to you that besides the duty of editing the *Encyclopædia*, and writing many of the articles, I have a correspondence to carry on with about an hundred different authors, who are writing for the work, and have often twenty or thirty letters lying by me unanswered. Even this labour, however, would not have prevented me from writing you, had I not, for these two years, been constantly determining to come to Jedburgh to see you. I have been hitherto unable to accomplish this journey, short as it is, though I am not without hopes of seeing you some time this harvest. . . . Your hygrometer of beechwood is very simple and ingenious ; but though it may be used for showing variations of humidity, yet it could scarcely be depended upon for measuring these variations. I am afraid you will find that the wood loses its power of expansion, and will not be so much affected by moisture a few months hence as it is at present. I consider the cold produced by evaporation as the only accurate measure of the degrees, and consequently of the humidity of the atmosphere. I have not yet got my achromatic telescope. There is no better method of polishing lenses than with pitch and flower of putty."

Most painful and harassing of all were the interrupted and broken friendships which were the result of this uncomfortable period, of which it is unnecessary at this lapse of time to say more, than that there were probably faults on both sides,—a commonplace, but true, solution of much that often appears inexplicable. In some instances, at least, the best and oldest of these ties were happily reunited. One bright circumstance shines like a sunbeam through the gloom connected with this literary undertaking. A request from Dr. Brewster to the Rev. Thomas Chalmers of Kilmany, to write the article CHRISTIANITY, turned the mind of the young and careless, though brilliant, divine, to study the truths of which he had then but a superficial knowledge, and ultimately proved the means of leading him to grasp them as a life-reality, with a force and power without which he could not have been the blessing to his country which he proved in after years. This was the beginning of a long and cordial friendship which only terminated with the death of Chalmers in 1847.

Whatever the pressure of work might be from the *Encyclopædia* and the *Edinburgh Journal*, Brewster never stinted or stayed in his own peculiar career. In 1811 he edited a new edition of Ferguson's *Astronomy*, to which he contributed an Introduction and twelve supplementary chapters. In 1812 he wrote the article BURNING-GLASSES for the *Encyclopædia*, containing a description of a polyzonal lens which he had invented the year before, when examining the experiments of Buffon. This lens, the source of much pain to himself, and yet, as many believe, of much blessing to his kind, will be described in another chapter. Veitch writes to him at this time :—

"INCHBONNY, *5th August* 1812.

"You have pointed out some very curious things concerning burning mirrors, and it would be of very great importance to put your devices into practice. I could wish that you would pay a little more attention to the fluid which you showed me, and find the true proportion of the curvatures of the two crown glasses, for I am convinced that it will answer better than the best flint glass that ever was made. I intend to make you a small reflector as a specimen of my workmanship; the tube will be about nine inches long, and two inches and a quarter diameter, or thereabouts.

"As for the comet, I do not know what to say about it; the first sight I got of it was on the 27th August 1811, and it was very nigh the star marked 26 on the shoulder of the Little Lion ; if I had lost sight of it three weeks after it had made its appearance, I would have concluded that its angle with the ecliptic would have been so great that it would have passed by the pole-star, for the only motion it had was that of latitude ;

but it lost that motion, and had little other motion but in longitude, till near its disappearance on the 25th of December, when it was in the 15th degree of Aquarius, with one degree of south declination; from the observations that I made on it, I was convinced that its path could neither be a straight line nor any regular course. Mr. Playfair told me that the astronomer at Glasgow had found its path corresponded very well with a parabola. It was as easy for him to make its path correspond with that curve, as to make one of Herschel's planetary nebula into a comet, which you will remember he published in the newspaper. On seeing the paper I sought diligently for it, and found it to be the planetary nebula discovered by Herschel on Feb. 1, 1785, vol. 75 of the *Transactions*, page 266.

" I must now remain, your sincere friend,

" JAMES VEITCH.

" The above was written before I received your letter. I am very happy that I have had it in my power to serve you at this time,—I did not make any delay in making the two glasses, as I am always very impatient about anything that I want, I thought you would be the same; they are scarcely so fine as I could have wished them, on account of the want of emery properly prepared, but I hope they will in some measure answer your purpose."

In 1813 Dr. Brewster sent his first paper to the Royal Society of London, on " Some Properties of Light," and in the same year he published a *Treatise on New Philosophical Instruments for various Purposes in the Arts and Sciences,*—a subject with which his early Inchbonny days had made him practically, as well as theoretically, acquainted. His health at last began to give way, or at least seriously required change of scene and complete relaxation. He therefore determined to take his first tour on the Continent. He had resided at Portobello for some time after his marriage, but had now settled in an unpretending "flat" in Duke Street, Edinburgh. His home had become the abode of little children. His eldest son James was born in 1812, and his second son, Charles Macpherson, a child of much love and much sorrow, was born in 1813. Mrs. Brewster could not, therefore, accompany her husband, and we see on this occasion much of the tender and beautiful romance of character which edged many of the dark clouds of his life with a silver light—his warm home affections being a characteristic through life, though often veiled and marred by a certain constitutional reserve, and other causes. He got a miniature portrait of his wife taken, as a travelling companion, and with a sad heart at the separation, set out on his first foreign journey. He writes to his wife, July 17, 1814 :—" I was in a very melancholy mood all yesterday, my dearest Juliet, at the prospect of such a long absence from you and my dear children ; but the rapid travelling, and the constant succession of new objects, have raised my spirits and habituated me in a slight degree to the first separation in our married life. My imagination, too, has been very kind to me. It has never allowed me to be for a moment absent from the only objects on which it delights to rest. I have seen every hour the scene that has been enlivening your little circle : our dear and intelligent little James kissing his sweet Charles, and their lovely mamma, like their guardian angel, watch-

ing her little charge, and teaching them to remember
their dear papa. Send me a lock of your hair, and
another of their's." Dr. Brewster continued a close
and minute correspondence with his wife, containing
everything that he could think of to amuse and
interest her in her quiet life and delicate health, even
to a sketch which he entitles, "Form of fashionable
bonnet," apparently combining the formation of a
helmet and a coal-scuttle! From these familiar letters
I extract the following, as interesting notices of the
great men of that now past generation, which show
the warm appreciation which he ever had of those
who had distinguished themselves in any branch of
science, and also the pleasant and honest surprise which
their recognition of him as a brother and peer caused
in his mind :—

"PORTLAND PLACE, *July* 23, 1814.

" . . . When we were about three miles to the north
of Slough, I observed in a carriage, which passed us
very rapidly, Mr. Watt and his lady, whom I was so
anxious to meet in London. The separation of the two
carriages was so rapid that I could not make myself
heard in attempting to stop the driver. When I reached
Slough I learned from Dr. Herschel that Mr. Watt had
spent three days with him, and had left him that morning. This was very mortifying to me, as I should
otherwise have had the pleasure of meeting these two
great men under circumstances of peculiar interest.
Dr. Herschel received me with the utmost warmth, and
begged that I would stay to dinner. He requested me
to present his compliments to La Place when I went to
Paris; and when I observed to him that I had only a
letter of introduction to M. Prony, and might not have

F

the honour of meeting such a man as La Place, he remarked that this was the same as a letter of introduction to all the French philosophers ; and that if I had no letters at all, my own name would be a sufficient introduction. This was obviously saying too much, but it was pleasant to receive such a compliment from such a great and venerable man as Dr. Herschel."

"PARIS, *August* 13, 1814.

" . . . Biot came to me very early in the forenoon, and repeated along with me the greater part of my experiments, leapt from his chair, clapped his hands, and constantly exclaimed, " *Oh que magnifique! oh que jolie!*" complimenting me in the true French style. I then accompanied Biot to the Library of the National Institute, where I had the good fortune to be introduced to Arago, an able astronomer, who has also made some fine optical discoveries.

" At three o'clock M. Biot and I set off for Arcueil, and on the road he promised to write the article MAGNETISM for the *Encyclopædia,* in which branch of science he has made several discoveries. When we arrived at the château of La Place, we were told by the servant that he was in the garden, and for nearly ten minutes we sought for him in vain among beautiful arbours and alleys of trees. He at last appeared, and gratified my curiosity, which had been wound up to the highest pitch. I was introduced to him by Biot, presented him with copies of my papers, and had a little conversation with him before dinner, in which he spoke of Mr. Playfair with kindness. La Place is a man below the middle size, of a fair complexion, and thin make. He is distant in his manner, speaks very little, and walks

with the stiffness of a senator. When he walks in his
grounds he carries a coarse stick about two feet higher
than himself, and wears a grey cloth cap resembling a
helmet. His hair was tied and powdered in the French
style, but in other respects he was dressed like an English gentleman. He was a great favourite of Buona-
parte, who loaded him with kindness, and took his son
along with him as one of his aides-de-camp. Our party
consisted of six, La Place and his son, his son-in-law,
M. Biot, and M. Poisson, a most distinguished mathematician, who has recently made some brilliant discoveries in electricity. Our dinner consisted of *soup,
fresh eggs, bouilli, roasted veal, mutton-chops, roasted
fowl, salad, French beans, honeycomb,* and a *kind of rice
potage,* besides a dessert of apricots, pears, cherries,
currants, and tarts, most of which dishes were served in
succession, and without any order whatever. I was
highly amused with the fresh eggs, and every person
during the whole of the dinner kept the same knife and
fork. After coffee we all set off to the house of the
celebrated chemist, M. Berthollet, for the purpose of
introducing me to him, and in order that he might see
my experiments on mother-of-pearl, which La Place
requested me to show him. He appeared, however, at
the end of a long vista of trees, accompanied by his wife,
on their way to La Place's château. A more homely pair
you never saw, and though very rich, they were dressed
little better than a decent Scotch farmer and his wife.
We then went into La Place's study, where M. Biot
explained at great length the different experiments
which I made. La Place was highly pleased with them,
and was very slow in believing that they were quite
correct."

"  . . . Went with M. Biot to a sitting of the National
Institute, and had an opportunity of seeing almost all
the distinguished philosophers in Paris.   There were
many strangers there, and several Englishmen of eminence, all of whom were seated at the backs of the
members, who were arranged round an oblong circular
table.   When the business commenced, M. Biot desired
me to follow him, and I was confounded when I saw
that he was taking me from the rest of the visitors into
the very centre of the circular table, where three chairs
were placed beside the President.   I was then introduced to the President, M. Lefevre Gineau, and ordered
to take my seat in that conspicuous situation.   In a
short time M. Biot left the hall, and I was left alone in
that solitary spot, wondering, along with all the other
visitors, why I had been placed there.   For a person of
my nerves this was sufficiently trying, but it did not
overpower me. . . . On Thursday I went to see M.
Arago, and the Observatory, which contains many
curious instruments, but few very good ones.   Arago is
a very interesting young man.   He was employed by
the French Government in making astronomical observations in Spain, and was thrown into prison and
cruelly treated.   He afterwards escaped to Africa, where
he travelled with a long beard as a Mussulman, and
through many hazards reached his native country.   He
has a lovely wife, and a son.   Being the particular
friend of the celebrated traveller, Baron Humboldt, the
Prussian Ambassador, he introduced me to him after
we had seen the Observatory.   I was very kindly
received by Humboldt, who was acquainted with my
experiments on Light, and I hope to see him frequently

before I leave Paris. He is a plain, frank man, and speaks English, French, Spanish, and German with equal fluency. I believe I have not mentioned to you that my book has been long known on the Continent by means of a most extensive analysis of it inserted in four successive numbers of the *Bibliothèque Britannique*, a work published at Geneva, and circulated in every part of Europe."

Besides correspondence, Brewster found time to keep a most minute and particular journal, which he seems to have had some thoughts of publishing, and for which his friend Williams, the eminent water-colour painter, drew six beautiful little sketches, which are now at Belleville. This journal, although now quite out of date, is extremely interesting, as showing the minute observation which was one of his characteristics even at that time, and probably one of the secrets of his success. Nothing escaped his quick eyes, which seemed to photograph on his mind every stone and crevice, every light and shadow, every window of a house, every colour of a landscape, every line or curve in pictures and statues. His journal is far more than a diary; it goes to the accuracy of hours and minutes. I shall only extract enough to show this characteristic minuteness and observation, and also some further interesting notices of French *savans*.

Dr. Brewster's appearance at this time was extremely prepossessing, although he had neither striking features nor commanding figure. His clusters of brown curling hair were often remarked, and the open, intellectual expression of his pale face, with the exceeding sweetness of his eyes. Although thirty-three, he was very

youthful-looking, so much so that, along with his extremely unassuming manners, the French philosophers
were quite puzzled, and it was probably at the very
meeting of the Institute described on another page that
they are recorded to have said, " What ! is that *boy* the
great Brewster ?"  The constitutional nervousness from
which he had long suffered showed itself principally at
this time in a degree of timidity which he made considerable efforts to overcome, thus alluded to amusingly in one of his home letters :—" I am trying as
hard as possible to get impudent.  I began this new
career by calling upon Mr. Sylvester, the chemist, at
Derby.  I tried it a second time at Oxford, and introduced myself to Dr. Robertson, Savilian Professor of
Astronomy, and I hope to have soon some other opportunities of showing off my new acquirements in this
way."

EXTRACTS FROM DIARY.

"PARIS, *Tuesday, August* 16, 1814.

" M. Biot and M. Cauchoix called upon me at two
o'clock, and showed me the ingenious instrument called
a Spherometer, invented by the latter, for measuring
the thickness of very thin plates.

" About half-past two o'clock I accompanied M. Biot
to the Institute, the ordinary meetings of which are held
in one of the apartments of the Library, in the Palais des
Beaux Arts.  The business transacted at this meeting
consisted in a report by M. Poisson, on some inventions that had been laid before the Institute by one
Muret ; a proposal by M. Legendre to alter the law
relative to the annual prize ; a communication by M.
Rossel ; an explanation of an improved circle, probably

by the inventor; and a long paper on iodine, by M. Gay-Lussac.

"As this meeting was well attended, I had an opportunity of seeing many of the most distinguished men in Paris; the principal members were—

"*Carnot.*—He resembled very much the picture of him which I have, but appears to be dissatisfied and discontented, and in bad health.

"*Legendre.*—A very tall and very thin man, with an expressive and intelligent countenance, white powdered hair, tied and curled above the ears.

"*Desmarets.*—An old, reverend-looking man. One of the old chemists.

"*Poisson.*—A young and active little man, with a sweet and expressive countenance.

"*Arago.*—Young (28), good-looking, dark, very pleasant and intelligent.

"*Monge.*—Below the middle size, stoops, has a full face, and white curled hair.

"*Lamarck.*—A good-looking old man, with a light coat and an embroidered waistcoat, little, and rather crooked.

"*Portal.*—A fine, reverend-looking old man, with a small face.

"*Gay-Lussac.*—A slender young man, a little marked on the face with the small-pox. Apparently a great enthusiast in chemistry.

"*Rossel.*—A little, thick, and active man.

"*Charles.*—An old man, intelligent face.

"*Burckhardt.*—A thin, pale, and slender young man.

"*Delambre.*—A little, oldish man, very yellow; a little marked with the small-pox.

"*Cuvier.*—Has rather the appearance of being self-sufficient ; is a little man, with a projecting brow and chin.

"*Huissard.*—A stout, and rather corpulent man. He sat on the left hand of the President; spectacles, and sallow. V.-President.

"*Prony.*—Not handsome ; large nose, intelligent and active.

"*Lefevre Gineau.*—Like Prony, so much so, that I took the one for the other. He is President of the class."

"PARIS, *Tuesday, August* 22, 1814.

"I went this morning to call upon M. Rochon, formerly the Abbé Rochon, a venerable and intelligent old man of seventy-three, who is well known to philosophers by his scientific works and inventions. He showed me his prismatic micrometer, a small instrument, with a level for measuring the inclination of lines to the horizon by the coincidence of two images, and his method of doubling the double refraction of Iceland crystal by extinguishing two of the images, and employing the two that are most remote. In the shop of the optician who works for him, he showed me a huge plate of glass, 6 feet in diameter, and 3 inches thick, which had been melted at the Gobelins in the time of Louis XVI., for the purpose of making a burning lens. He expects that it will soon be ground, the operation being already begun. The instrument mentioned above for measuring the inclination of lines to the horizon was first suggested and described in my *Treatise on Instruments*, the only difference between the two being in the way of forming the double images. M. Arago having mentioned to me that M. Rochon had discovered before me the double dispersive power of calcarean spar, I

replied that he had given merely the two *dispersions*, not
the two *dispersive powers* of that crystal.   M. Arago
assured me that this was not the case, and I of course
applied for information to M. Rochon himself.   He
showed me the table containing the results of his
experiments, which was exactly what is given in Cavallo's *Natural Philosophy*, to which I have alluded in
my book.   I explained to M. Rochon that his results
were merely the dispersions, and he admitted that I
was perfectly correct.   The slightest examination of his
table, indeed, is a sufficient proof that this is the
case."

Brewster and his travelling party left Paris *en voiture*
on August 28th.   He stayed three days only at Geneva,
where, however, he made acquaintances so agreeable
and so congenial that they were never forgotten in
after life, Professor Prevost, M. Pictet and his family,
amongst the number.   Sir Humphry and Lady Davy
he also met frequently.   An expedition to Ferney interested him much ; in Voltaire's bedroom he saw an
engraving of Newton, which he long afterwards described.   The short tour through France and Switzerland was concluded by their arrival in England on
September 28th, the journal being continued by hour
and minute.   Thus we have—
" *Sept.* 10, *Sat.* 9.25.—Cross the wooden bridge of
St. Pelissier, which is now very good and safe, and
from which there is a fine view of the Arve, Mont
Blanc towering above it.   We then ascend a most
dreadful road over unbroken masses of rock.   I observed several fine examples of the scoops and grooves
which Sir James Hall observed upon Corstorphine Hill.

They stretch in the direction of S.E., which is the direction of the valley. . . .

"9.45.—Most beautiful scoops on the right hand, and ruts in the rock; many of the hollows are finely smoothed out.

"9.48.—The ruts and scoops are here most distinct, and more perfect than any of those seen at Corstorphine Hill.

"9.50.—Enter the Valley of Chamounix, and observe scoops below the road near the Arve.

"9.58.—A fine hollow on the right, where the mountains retire, in which is situated the village of Chavan, on the banks of a brook; the Aiguilles of Mont Blanc are now seen. . . .

"10.50.—Enormous blocks of stone, not rounded, appear here; a little wheat is grown in this quarter.

"11.—Descend and walk to the Glacier de Boisson, which is very fine; the peaks of ice are extremely grand, and have a fine blue colour in the crevices; at the side of the glacier numerous trees are crushed to pieces; large blocks of granite are suspended on the flanks of Mont Blanc, and the earth is turned up in such a manner as if some great convulsion had taken place. The plateau of the glacier is in no respect remarkable; huge blocks of granite are lying upon the ice, by the descent of which they are transported to a lower level. . . .

"3.25.—Reach the summit of Montanvert, and rest in the house erected by M. Felix Deporte, once the French resident at Geneva, for this purpose. There was a fine wooden fire blazing, and we were supplied with bread, milk, and cheese. The names of De la Saussure, Dolomier, La Lande, and Pictet are painted on the walls.

" 4.3.—Set out to see the glaciers, and descend a hill till we reach the Mer de Glace. It is like the waves of the sea, as if they had been fixed by sudden congelation. Where the ice is most perfect, which is on the sides of the deep crevices, the colour is a fine blue. There is an appearance of a vertical stratification in the icy masses stretching in the direction of the valley in which the glacier lies. We passed a huge granite block, about 25 feet high, resting upon the ice. It descends continually. The noise of the waters rushing below is very fine, and the sound of large stones or masses of ice tumbling into the crevices continually remind the spectator that his situation upon the summit of the frozen waves is not exempt from danger. Towards the edge of the Mer de Glace the ice is covered with pounded granite, and huge masses of that stone mark the boundary between the ice and the mountains. In these places where the ice is covered with sand, it has the appearance of being perfectly black, like the darkest Cairngorm, while in other places the perfect ice is green. Upon breaking this apparently black ice it is perfectly transparent, and remarkably pure and hard.

" The surface of the glaciers exhibit also the appearance of veins exactly like rocks of stone. . . .

" 4.56.—When we had reached the bottom we were accosted in English by an old man, who was one of the two Cretins described by Saussure, and exhibited in England. I recollected having seen him in Jedburgh many years ago along with his brother. He is very poor, and draws a small subsistence from the generosity of the English. . . .

" *Sept.* 12, 4.15.—Reach Vevay, where we stop for the night. It is a large town, with many excellent houses, but the streets are narrow, and without foot-paths. In the evening I walked along the banks of the lake in a fine mall of trees, from which there was a charming view of the sun setting behind the ridge of Jura. He had just descended below the horizon, and left a fine glow of purely yellow light along the whole of the western sky. Above was a warm glow of red, and the whole extent of the lake towards the west was of a lovely purple colour. As the sun descended the yellow gradually deepened into orange, and the purple glow upon the lake became more faint. This lovely scene formed a grand contrast with the dark gloom which was thrown over the Eastern Alps, and the blackness of the part of the lake which intervened. . . .

" *Sept.* 15, 11.27.—Passed some cottages ; the country beautiful, and wooded. After passing through a very pretty country we reach the village of Lechelin, with a church, and many thatched houses with roofs extremely steep. The women most extraordinarily dressed, with a circle of wrought horse-hair like gauze sticking up over their heads, and their hair in two large plaits behind, with strings tied to it, and reaching to their feet. . . .

" *Sept.* 16, 4.25.—Reach the Lutschine, a large and rapid river formed by the Black Lutschine, which runs through Grindelwald, and the White Lutschine, which flows through Lauterbrunnen. The rocks on the right are so high that they are covered with snow. On the left is a lofty mount, with singular convolutions on the strata.

" 4.55.—Most extraordinary rock on the right, with trees on its perpendicular face like men walking up. These are called the rocks of Eisenflue ; they are 900 feet high, and have a village of the same name on the top.

" 5.25.—Most extraordinary convolutions on the left towards the root of the rock; they are really double convolutions included in a single one. . . .

" 6.35.—Turn back to the inn; the valley and its precipitous flanks were almost wholly in total darkness, while the red twilight shed a bright hue over the Jungfrau, the Breithorn, and the Lauterhorn. The whole appeared as if we were looking out of a dark room into a higher region. This appearance was still more striking at a later hour, when Saturn was seen over the Jungfrau, and when the lights in the cottages appeared like so many stars in the dark declivities of the valley. The blue and white streaks on the calcareous rock had a very singular appearance."

## CHAPTER VII.

### NOTES OF LIFE FROM 1814 to 1824.

Wisdom's self
Oft seeks a sweet retired solitude ;
Where with her best nurse, Contemplation,
She plumes her feathers, and lets grow her wings,
That in the various bustle of resort
Were all too ruffled, and sometimes impaired.

MILTON.

BREWSTER'S continental holiday was not much of a holiday to those at home. His door was besieged with letters, messengers, and printers' devils—the *Encyclo-pædia* irregularities grew worse and worse, and many an application was made for the editor's speedy return to the scene of action, to regulate contributors and publishers. When he did return, it was therefore to double tides of work, which he plunged into with unabated ardour. He sent to the Royal Society of London a series of nine papers, spreading through several years, most of them on his favourite subject of the Polarization of Light, in which demesne he had already made great and original discoveries, contributing also many papers to the Royal Society of Edinburgh on this and other subjects. In 1815 he became a Fellow of the Royal Society of London, which afterwards bestowed upon him the Copley Medal; three years later, the Rumford; and subsequently six of the Royal Medals, in each case for fresh discoveries in Light. In this year he was invited to conduct the class of Natural Philosophy during the absence of Professor Playfair, who was abroad, at the especial request of the Corporation of Edinburgh, as well as by the wish of the Professor himself. He did not undertake this task, however, as it was not suited to his taste, and would have robbed him of the scanty time he possessed for experiments and discoveries—ever the work which he fell back upon, in the intervals of a life so busy, that few would have discovered in it interstices of leisure sufficient to make any use of. In 1816 the French Institute awarded half of the prize of three thousand francs given for the two most important discoveries in physical science made in Europe to the " boy," who had sat in their most distinguished seat, and who, even in the two years that had passed since then, had made fresh and important progress.

In this year Dr. Brewster invented the Kaleidoscope, which, though of little practical advantage, spread his name far and near, from schoolboy to statesman, from peasant to philosopher, more surely and lastingly than his many noble and useful inventions. This beautiful little toy, with its marvellous witcheries of light and colour, spread over Europe and America with a *furor* which is now scarcely credible. Although he took out a patent, yet, as it often has happened in this country, the invention was quickly pirated, and thousands of pounds of profit went into other pockets than those of the inventor, who never realized a farthing by it. Two years after, Dr. Brewster went up to London, in the vain hope of rectifying this mismanaged business, and making it profitable in a way which would have been very useful with the claims of an increasing family pressing on him, and only the precarious profession of literature to fall back upon. The following letters give an amusing

idea of the ferment on the subject which prevailed both
in England and Scotland, and show also the misman-
agement which had taken place. He writes to his wife
from Sheffield, May 17, 1818 :—

"We arrived here this forenoon from Leeds, and
have been obliged to remain all day, in consequence of
there being neither coach nor mail for Birmingham after
our arrival. We have, however, spent our day very
agreeably in visiting the principal manufactories, with
which we were much delighted and entertained. We
were introduced to most of them by Mr. Cutt, the
partner in the house of Cam and Cutt, who have undertaken to manufacture the kaleidoscopes for Mr. Ruthven. They have agreed to make and sell the instruments under my patent on the same terms as Mr.
Carpenter, provided I get his permission to allow
them to be employed. This I must do, as he cannot
possibly supply the demand. On my arrival at the
Tontine Hotel here, the first sight that displayed itself
was a pair of kaleidoscopes in two tubes (most deplorable instruments) lying on the chimney-piece. The
waiter told us that they were invented by a doctor in
London, who had got a patent for them,—that, by some
variations, the tinmen had invaded the patent, and that
the said doctor was trying to find them out and prose-
cute them! The Sheffield newspaper lying on the
table contained a flattering paragraph about the same
instrument; and when I called on Mr. Cam, I saw
lying on his table a kaleidoscope, most beautiful on the
outside, but deplorable within. I am wearying sadly
to see you and the four dear boys, and shall make my
stay in London as short as I possibly can. I have
often repented leaving home in search of what could

not make us happier; but it is perhaps right that I have done this, and if any good comes of it, I shall have the more reason to congratulate myself upon having taken this step."

And a week or two later he writes:—

"LONDON, *May* 1818.

"I dine to-morrow with the Royal Society Club, and in the evening I undergo the ceremony of being admitted a member, which is a more formal business than I had supposed. I called yesterday at Sir Joseph Banks', and met Sir Everard Home, and other wise men there. Both of these gentlemen assured me that had I managed my patent rightly, I would have made one hundred thousand pounds by it! This is the universal opinion, and therefore the mortification is very great. You can form no conception of the effect which the instrument excited in London; all that you have heard falls infinitely short of the reality. No book and no instrument in the memory of man ever produced such a singular effect. They are exhibited publicly on the streets for a penny, and I had the pleasure of paying this sum yesterday; these are about two feet long and a foot wide. Infants are seen carrying them in their hands, the coachmen on their boxes are busy using them, and thousands of poor people make their bread by making and selling them."

Mrs. Brewster, on the other hand, gives the experience of Edinburgh on the same subject. She wrote to her husband:—

"*Friday, May* 22, 1818.

"MY DEAREST HUSBAND,—I was very much disappointed that I did not get your promised *long* letter to-day, and so was poor Mr. Ruthven, who flattered

G

himself that you might naturally have made some mention of Carpenter's extraordinary conduct in not supplying him with any instruments, nor even sending him a line to account or apologize for the same. You never saw a person in such real distress as Ruthven is—the public are becoming impatient and clamorous now at the delay, and he has orders to an amount that is prodigious. One person offers him the money for 150 to send abroad in ten days. The ship sails then, and he will not take the instruments after that period. Mr. Ruthven has come to the resolution of setting off to-morrow for Sheffield and Birmingham. People insist on leaving their money in advance in order to secure their chance, and from six in the morning till six at night his room is beset with people. They cannot understand how completely mismanaged it has been, and that the capital of Scotland, and your place of residence, should not contain a single kaleidoscope for sale for the last eight days! I am sorry to vex you, but if you saw Mr. Ruthven's face you would not wonder at my urgency! . . . I told him about the cheap instruments, but he says that here he can sell any number of the pound ones, and thinks it quite useless to begin any lower at present; indeed he is not in a train for anything just now, for he is evidently ill with extreme vexation and fatigue. Patrick was at his house this evening at half-past eight o'clock, and found it full of people all wanting kaleidoscopes, and this has been the case for the last ten days.

"*Saturday.*—Mrs. Ruthven has now to bear the toil of the people wanting instruments, and she has sent to beg that I will desire you to send down kaleidoscopes of all sorts. You would make a fortune at the present period, but this delay is worse than all the piracies that ever were attempted. Ruthven could sell a hundred per day, and at Glasgow they are quite wild, and at Montrose the same, and at Paisley, and, in short, everywhere, but if this *fever* is to be balked much longer, I am afraid people will be quite disgusted. *Do* something about the kaleidoscopes, or Mr. Ruthven will lose his wits!"

The *Edinburgh Magazine* had now taken a different form, and under the name of the *Edinburgh Philosophical Journal*, Brewster edited it in conjunction with Professor Jameson, the eminent Scotch mineralogist, and afterwards alone, as the *Edinburgh Journal of Science*, of which he published sixteen volumes, which were remarkable for freshness and variety, and for the eminence of the contributors. Editorial labours, indeed, occupied him for a large portion of his life, and he contributed many original articles to the works under his charge. In 1821 he took an active part in founding the Royal Scottish Society of Arts, of which he was named " Director,"—the original intention being to establish permanent Schools of Arts in Edinburgh and other large towns. This idea was afterwards abandoned, and although for some years Dr. Brewster continued to interest himself in the affairs of the Society, his office in it was somewhat an anomalous one, and after 1827 his name only occurs in connexion with its meetings in 1849, when he contributed a paper upon Stereoscopes. In 1820 he became a Member of the Association of Civil Engineers in London, and in 1822 of the Royal Irish Academy of Arts and Sciences. In the latter year he edited a translation of Legendre's *Geometry*, and also four volumes of Professor Robison's *Essays on*

*Various Subjects of Mechanical Philosophy,* and in 1823 he edited Euler's *Letters to a German Princess,* with notes, and a life of the author.

Through all this network of interests and occupation, it is interesting to find the long-loved telescopes and the Inchbonny correspondence still holding their own place, although the letters were necessarily less frequent as the press of other work got more burdensome. We have the following request for experiments on different telescopes, to the man of more leisure, but not less mental activity, and his response :—

"EDINR., 13 HOPE STREET, *April* 25, 1816.

" I am glad you are occupied in determining the latitude and longitude of Jedburgh. I wish you would try some experiments on the difference between the Gregorian and Cassegrainian telescopes. It appears now quite certain that whenever the rays have a positive focus, a great quantity of light is lost by their collision. Hence the great superiority of concave eye-glasses above convex ones. Mr. Playfair had unluckily mislaid your method of casting and grinding specula, but he has now found it, and it will soon be printed in our *Transactions.*"

"INCHBONNY, 31*st May* 1816.

" I have made many experiments on the Cassegrainian telescopes, and I find when the telescope is short the Cassegrainian has the advantage of the Gregorian ones, perhaps more than one-third of the whole magnifying power; but when I tried it on my 5-feet Gregorian reflector, I could discern no difference between the convex and concave small speculums on Saturn, although the speculums were both the same radius, and equally distinct when tried on the double stars. I have made a Gregorian telescope, 34 inches focus, diameter of the metal 5½ inches. It is first a Gregorian telescope with two set of eye-glasses magnifying 78 and 150, and then a Cassegrainian telescope with magnifying powers of 173 and 300; with the power of 173 Cassegrainian, it is equally bright and distinct as the power of 150 Gregorian. I see Jupiter's belts exceedingly well, and the disc of Jupiter, round and well defined, with the last power of 300. I sold this telescope the other day at twenty guineas.

" It appears from your letter that Mr. Playfair intends to publish my paper in the *Transactions*. You will be so good as to desire him to suppress anything that he sees superfluous in it, especially that part which estimates what we might expect from Herschel's 40-feet reflector,—it is far from my wish to undervalue any man's performance. You will get fine views of Jupiter at present; he is a fine object for trying the goodness of a telescope on, as the least imperfection in the figure will easily be perceived by a false glare of light blazing about the planet. Your Dictionary (*Encyclopædia*) is every day meeting with new applause, and I hope will continue so until it be finished. You have never sent your little telescope that I may put it right for you. I saw your picture at Dryburgh the other day, placed on the side of the window beside James Watt, just as you enter his Lordship's room. The painter has made your eyes as big as William Wilson's. In other respects it is not unlike you."

Copies and occasionally proofs of his own and other scientific articles were always sent to Inchbonny, where appreciation and interest ever awaited them. Mr.

Veitch's observations and experiments are always
noticed as exciting his old friend's grateful interest, and
of some he says, "I shall print them in next number of
the Journal,"—while he looks to Mr. Veitch "for much
valuable assistance in the article GRINDING for the
*Encyclopædia.*" Lenses, diagrams, and specimens are
interchanged and commented on. On one occasion
Brewster sends "a fragment of garnet, but cannot see
how you can get it cut at Jedburgh." Genius laughs at
impossibilities, however, for some years later Dr. Brewster,
in his "Optics" in Lardner's *Encyclopædia*, says of
this same piece of garnet, "Mr. Veitch of Inchbonny
has likewise executed some admirable garnet lenses out
of a Greenland specimen of that mineral given me by
Sir Charles Giesecké." Then, in 1821, Brewster has
the pleasure of forwarding to his friend an order for
one of his best telescopes, a Gregorian reflector, from
Professor Schumacher of Altona, Hamburgh. It was,
when completed, 2 feet 8 in focal length, and 5 inches
aperture, and proved a very fine instrument. A telescope of the same description was made for the Earl of
Minto, and another is in the possession of his son, the
Rev. Dr. Veitch, at Merchiston, near Edinburgh. Dr.
Brewster about this time took a hasty run to Kelso to
see his old friend the Rev. Mr. Lundie, and to Edgerston
and Jedburgh. His visit being unexpected, Mr. Lundie
was absent, and his sensitive temperament made the
whole expedition one of extreme and impulsive distress.
He thus writes to his wife :—

"KELSO, *Saturday morning*, 1820.

"I set off for Jedburgh early next morning. The
first person I met in the street was James Veitch, and
having learned from him that Dr. and Miss Somerville

were from home, I set off with him to his house at
Inchbonny. I found upon inquiry that scarcely any of
my old friends were in existence. Most of them were
dead, some deranged, and others ruined, so that I had
really nobody to see or to inquire after. In short, notwithstanding the increased beauty of the picturesque
scenery round the town, I found every object associated
with the most distressing recollections, and would have
been happy had I been able to place my mind in its
former state of ignorance. The place was to me like a
sepulchre, where some of the limbs of its tenants had
scarcely expired. I had resolved, after seeing all Mr.
Veitch's curiosities, to go to Edgerston, and just as I
was returning to Jedburgh for that purpose, I met Mrs.
Rutherfurd, and told her my intention. I accordingly
went there to dinner, and stayed all night, and such
was the state of agony in which I found myself when
I had time to think of all the painful recollections
which my visit to Jedburgh had awakened, that I resolved to return to Kelso by a shorter road, and leave
all my visits in Jedburgh for another year. This was
very selfish, but I saw no reason for incurring any
further pain. . . . On Friday (yesterday), Mr. Lundie
arrived; and Mrs. Lundie, who is one of the most charming and intelligent women I ever knew,[1] invited Dr.
and Miss Somerville and General and Mrs. Elliot to

[1] This lady afterwards married Dr. Duncan of Ruthwell, another friend
of Dr. Brewster, the author of the *Sacred Philosophy of the Seasons*, and
the *Cottage Fireside*. She was the mother of the beautiful and early
sainted Mary Lundie Duncan, and the authoress of her charming *Memoir*
and other interesting works. My father was much attached to all this
circle, and in 1863 wrote thus to Mrs. Lundie Duncan,—" Your letter has
indeed called up the remembrances of early and dear friends, and of the
many happy days spent in their society ; but in proportion to the distance
of these days is the shortness of the time when we shall rejoin them."

dine with them. I have just come from breakfast at
Rosebank. Lundie and I dine to-day at Fleurs Castle,
where I stay all night, and set off on Sunday morning
to church at Dryburgh Abbey, where I remain all Sunday night, and go to Stow on Monday by the Jedburgh
coach, which passes Lord Buchan's."

One of the good results of Brewster's wonderful art
of "making time" was, that he still always contrived to
keep some space for exercise and relaxation, without
which he never could have lived such a life of labour
with so few traces of overwork. Of this quality he
thus wrote to Miss Edgeworth :—" I am one of those
ill-organized people who cannot work by rule. I must
be in a fit either of unremitting labour or of absolute
relaxation, taking no amusement in the one paroxysm,
and doing no business in the other." His intense love
of the country, and of all country pursuits, contributed
greatly to his health both of body and mind. His early
love of shooting had not deserted the philosopher, although his skill does not seem to have increased greatly
since the days of the Auld Wood and the Ana. We
find a characteristic sentence in a letter,—" Determined
to shoot better than last Saturday," followed by a triumphant announcement, " shot seven brace of grouse."
He had now full scope for this amusement, as the
beautiful and well-stocked moors of Belleville, his
brother-in-law's residence in the Highlands, were open
to him, and he proceeded thither almost every shooting
season for many years. His wife's health was sometimes too delicate to allow of her accompanying him on
these visits, and we have the following graphic description of the pleasures and pains of the moors :—

"BELLEVILLE, *August* 13, 1820.

" Take great care of yourself, my dearest Juliet, and do not let any of the little vexations of this world give you the least uneasiness. If I had had you among the hills, where not a trace of worldly concerns or worldly localities are to be seen, you would have at least resolved to trouble yourself little about the miserable anxieties of the valleys. I know of nothing more grand and impressive than a day's sojourning among the huge and unchangeable masses which compose this heathy wilderness. We can almost believe that we are without a home, without a country, without a family, when we see none of the circumstances which are usually associated with these distant ties, and have before our eyes only heaven and earth, and before our minds only the great Author of nature. I expect to enjoy these sensations still more powerfully to-morrow among the ravines of the Dulnain."

Another year he writes :—" The 12th of August here brought along with it more than its usual allowance of disappointment. John Grant [of Rothiemurchus] and I prepared ourselves at five in the morning for the labours of the day, but we had scarcely finished our breakfast before torrents of rain fell from every part of the horizon. The weary four hours which preceded the appearance of the family were spent in two perusals of Rose's book on Monkeys—in twenty or thirty games at battledore and shuttlecock, and in many appeals to the barometer, and to the faithless skies. About eleven o'clock there was a sufficient intermission of the rain to induce us to take the hill, especially as there was a party to dinner, which required some game, on the 12th of August. The first bird we saw I shot, but at the same instant broke one of my detonating locks, so that during the rest of the day I was confined to one barrel. After I had shot five brace, and John Grant four, during which time we were completely drenched, and no longer able to work our guns, we left the hills about three o'clock. The birds are by no means numerous; we never saw a larger covey than five during the day. Yesterday, the 13th, we were again equipped for the hills, and the horses waiting for us; but the day was again unfavourable, and I did not think it prudent to expose myself a second time to the elements. If to-morrow is at all favourable, I intend to have a complete day of it alone with James Donaldson, and as my gun is mended, I expect considerable success. If to-morrow is a bad day, as it threatens to be, I intend to go to Kinrara, to pay my respects to Lady Huntly, who arrived yesterday." The acquaintance alluded to here with "Elizabeth, last Duchess of Gordon," then Marchioness of Huntly, ripened into an interesting correspondence, as was so often the case with Dr. Brewster; it was chiefly upon mineralogy, in which Lady Huntly was most intelligently interested, and she was able also to understand and appreciate Dr. Brewster's optical experiments. He was at this time greatly occupied with meteorology, taking much trouble to get a correct series of observations made throughout Scotland, and Lady Huntly got observations taken for this purpose both at Huntly Lodge and Gordon Castle.

Two other children, David Edward, born in 1815, and Henry Craigie, born in 1816, had been added to the home circle, which now gathered in a commodious house, No. 10 Coates Crescent, then a new and almost rural part of Edinburgh. One summer Dr. Brewster took his family for a happy summer and autumn's residence to Venlaw, a pretty place on the Tweed, near Peebles. Nowhere did the hard-worked man relax the tight-strung bow more completely than in this retreat. With his wife mounted on a beautiful little pony, called the Black Dwarf, and his two eldest boys, he thoroughly explored the beautiful scenery around. Congenial friends and neighbours crowded around them, and both husband and wife looked back to, and often spoke of, this time as one of the happiest of their lives. With one of their near neighbours[1] a life-long friendship was commenced. A member of this family writes to me that my father and her father used to take at this time long walks together, which were a source of the keenest mutual enjoyment, not by any means from the least interchange of scientific or literary ideas, but simply from the intense sympathetic love of the beauties of nature which was prominent in both.

The happy summer at Venlaw doubtless inclined Dr. and Mrs. Brewster's mind to pitch some little country tent of their own in the picturesque Lowlands of Scotland. There was a beautiful little property near Jedburgh, named Allars, which they were much inclined to purchase, but with a characteristic timidity and strange foreboding, Dr. Brewster gave, as one of his principal objections, the dangers to his children from the vicinity of a mill-stream which ran immediately outside the gates. A place near St. Boswell's seems also to have been on the *tapis*, for Mr. Veitch writes, not diminishing its disadvantages:—" I was look-

[1] Robert Montgomery, Esq., who was then residing for a time at Kailzye.

ing at the house at St. Boswell's, and I think it is the most lonely place I ever saw, and probably much haunted by evil spirits. There is a deep ravine at the back of the house, and at the brae-foot the Tweed, with a whirlpool 20 or 30 feet deep; if one of the boys were to slip in, he would never come out again. If you had a house near Jeddart you would be far better, and against summer you and I would go to Stewartfield Wood and shoot 'craws,' and have fine fun!" At last a pretty site was fixed on, just opposite Melrose Abbey and the Eildon Hills, with the sparkling Tweed rolling between, yet, as was supposed, at a safe distance, and purchasing a small farm, Dr. Brewster built an unpretending dwelling—half cottage, half villa, which he named Allerly, partly from his pleasant recollections of Allars, with its green alders, and also true still to early memories, from the Allerly Well, a gushing spring between Jedburgh and Inchbonny, where he had as a boy so often quenched his thirst. The superintendence of the building, and the laying out the garden and grounds, every tree being planted with his own hands, took him often from Edinburgh to Roxburghshire, and proved a most healthy relaxation.

In October 1823 the birth of their only daughter nearly brought desolation into the home in Coates Crescent. The mother's life was for a time despaired of, and it was not, therefore, till the spring of 1824, that the family, in recovered health and spirits, moved to the little home beneath the shadow of the Gattonside hills, although for some years they continued to reside in Coates Crescent in winter.

# CHAPTER VIII.

## MISS EDGEWORTH—JUNIUS.

FRIENDSHIP is no plant of hasty growth,
Though planted in Esteem's deep fixed soil;
The gradual culture of kind intercourse
Must bring it to perfection.
                            JOANNA BAILLIE.

As govern'd well or ill, States sink or rise ;
State ministers, as upright or corrupt,
Are balm or poison in a nation's veins ;
Health or distemper ; hasten or retard
The period of her pride, her day of doom.
                            YOUNG.

IN 1823 Miss Edgeworth visited Scotland, was received as a most honoured guest at Abbotsford, and
remained some time in Edinburgh.   Here she and
Brewster commenced a most cordial friendship, extending also to other members of the family, as through his
introduction she visited the Macphersons at Belleville,
and enjoyed exceedingly her Highland experiences.
Miss Edgeworth and Dr. Brewster carried on for many
years a close and lively correspondence, in which he
found time to communicate many of the subjects which
engrossed his attention from time to time.   With the
exception of his correspondence with Mr. Veitch, and
his regular home letters in any time of absence, this
seems to have been almost the only long voluntary
correspondence of general interest which he ever entered
into *con amore*.   With so much necessary writing to
accomplish, he generally looked upon ordinary letter-

writing as an unpleasant task and a waste of precious
time.

From the following letter we find that it was about
this time that a subject was brought prominently before
Brewster, which at intervals much engrossed his attention during the whole of his life. In his student days
he had read the letters of that mighty unknown man
who levelled the weapons of the fiercest denunciation
and the most burning eloquence against the corruption
of States and statesmen. The anonymous scabbard of
"Junius" was all that stood between him and condign
punishment; and frail as the interposition seemed, the
preservation of his great secret is perhaps as wonderful
as the daring and grandeur of his sarcasm. As we have
seen hinted by Dr. Andrew Thomson, Brewster had
perhaps somewhat profited by his perusal of the wit
and satire of Junius, but it was probably not till the
incident mentioned in his letter that the thought crossed
him that he might discover the secret so marvellously
preserved :—

                        "EDINR., 10 COATES CRESCENT,
                              *Dec.* 8, 1823.

" MY DEAR MISS EDGEWORTH,—I fear that you have
thought me negligent in allowing your kind and welcome letter to remain so long unacknowledged, but I
am not without hope that you have imagined some
apology for me. I had scarcely read your letter when
my poor wife was taken so ill that I have hardly been
sensible that there were any persons in the world but
ourselves. She has now recovered so completely, however, by adding a little girl to our group, that I have
again begun my commerce with the world by discharging one of the most agreeable of its duties. . . .

I had the pleasure of seeing a good deal of your friend Dr. Brinkley. Among the great men of the present day, so fatally characterized by want of principle of every kind, it is truly refreshing to meet with such a man. I was no less delighted with his lady and family, and ever since your visit and theirs to our metropolis, I have indulged in many an air-built scheme of treading upon your green island. If the spider goddess would only throw an iron web over the channel, or some teredo of an engineer cut out a tunnel beneath, I should then expect to realize these happy visions; but that boisterous Irish sea of yours is a sad enemy to those who, like me, cannot float in tranquillity upon salt water.

" When you were in Edinburgh I was anxious to have shown you some curious documents which I have collected respecting the author of 'Junius,' not only because I know that you would be much interested in them, but because I expected that you would give me some help in the inquiry. The subject is out of my line of study, but as it has crossed my path, I feel an obligation to pursue it. In looking over some old papers of the late Mr. Macpherson of Belleville, I found a packet of hurriedly-written notes on East-Indian affairs, signed 'Lachlan Macleane'"—[both were confidential agents of the Nabob of Arcot; Mr. Macpherson was one of the Government writers against " Junius," as " Scævola," and various other signatures]. " The first that I read had the following sentence :—' The feelings of the man are not fine, but he must be chafed into sensation.' A passage so beautiful and so like Junius, that upon no other evidence I supposed that I had found out the great secret. On a little reflection I recollected a story in Galt's *Life of West*, relative to 'Junius,' and was impressed with the belief that the name of Macleane was used in it. This I found to be the case, and I have now accumulated such a body of evidence that Henry Mackenzie, and many persons here who are good judges of evidence, consider the point as nearly made out.

" Macleane was an Irishman, and his father, who was a Scotchman, was minister of Rachry, near Belfast. He studied medicine at Edinburgh, and went out in Otway's regiment to Canada as an army surgeon. His talents were of that kind that he became private secretary to General Monckton, commanding in the West Indies, and he afterwards practised as a surgeon in Philadelphia. An intrigue with one of his patients drove him from that city, and in crossing the Atlantic in the same ship with Colonel Barry, this gentleman became acquainted with his great talents and recommended him to Lord Shelburne (the late Marquis of Lansdowne), as Under Secretary of State. In this capacity he acted for some time, till he and Lord Shelburne were turned out of office by the Duke of Grafton's ministry. Macleane lived in London during all the time that Junius wrote. He was a great gambler and dabbler in the funds, and made much money. It was well known that he moved in the first political circles ; and though he and his party were out of office, he had a splendid appointment created for him in India, namely, that of Commissary General in Bengal, with the emoluments of a Junior Counsellor. This appointment was conferred upon this ex-secretary three months after Junius ceased to write. On his return from India in 1780, he was lost with all his papers in a vessel ['a crazy vessel, commanded by a crazy captain,' as he described it before sailing], which was never heard of. His talents were of the first order, but they were not generally recognised, as he stammered greatly, and therefore never tried to shine in conversation.

"I have learned that he had two sisters resident in the North of Ireland, but I have not been able to find their place of residence. It is very probable that they could communicate some important facts. . . . Having thus afflicted you with a long story, I must not try your patience any longer. A packet came here a few days ago for Mrs. Macpherson, which, from its form and specific gravity, I judged to be your tabinet. I sent it off immediately, and I daresay you will in a few days hear of its safe arrival.

"We are all impatient here for the appearance of St. Ronan's Well, which is said to be excellent. I was highly amused a few days ago at the anniversary dinner of the Antiquarian Society, to see Sir Walter joining in all the honours, when the 'Author of Waverley' was given as a toast. He is in great health and spirits. Mrs. B. joins me in kindest compliments to you and Miss Harriet and Miss Sophia, and I am, my dear Miss Edgeworth, ever most faithfully yours,

"D. BREWSTER."

Brewster turned his energies most characteristically at this and other times towards unearthing "the mighty boar of the forest," as Burke termed Junius, sparing no pains by correspondence in Scotland, England, Ireland, and America to get new facts bearing on the weird authorship, Colonel Lachlan Macleane being,

H

however, his favourite claimant to the doubtful distinction. He thus collected much fresh and important evidence in favour of Macleane; his birth as an Irishman, yet his intimate connexion with, and knowledge of Scotland, the frequent terse and technical medical references scattered throughout Junius, the similarity of handwriting, the identity of dates between the first publication of the letters and Macleane, with his patron, going out of office; and again, the sudden stoppage of the warfare just before Macleane got the sop of a lucrative appointment, bestowed under very unlikely circumstances, and the appearance of his supposed portrait in a curious contemporary print found in Dublin, called " the Tripartite Junius." These, with many other curious particulars, Brewster gave to the reading world in two most interesting and clearly written articles in the *North British Review*, in which he submitted the arguments in favour of Sir Philip Francis, Lord George Sackville, Colonel Barré, Lord Temple, and Lord Lyttleton, to a most searching and critical investigation. His own arguments for Macleane, if they did not carry entire conviction to the public, certainly created a strong rival to the claims of Sir Philip Francis, the only other formidable competitor. To avoid recurrence to the subject, I may mention here that the Memoirs of Sir Philip Francis, by Mr. Parker and Mr. Merivale, published in 1867, was the last secular book which my father read. It occupied him greatly; he particularly requested his daughter-in-law to read it, and he said to her that though " not convinced," it had " greatly staggered " his previous firmly held views of the authorship of Junius.

Brewster's admiration of the composition, intellect, and eloquent denunciation of wickedness in high places, found to perfection in the writings of Junius, was intense. The following extracts from the *North British Review*, while offering an eloquent apology for Junius, may be taken as a recognition both of his vices and his virtues :—

" There are infirmities, however,—there are even vices, which shrink from the public gaze, and which neither invite our imitation nor demand our rebuke. Charity throws her veil over insulated immoralities, into which great and good men may be occasionally betrayed, and which accident or malignity may have placed before the public eye. When remorse or shame pursue the offender, public censure may well be spared. Vice has no attractive phase when the culprit is seen in sackcloth or in tears. But when licentiousness casts its glare from a throne, or sparkles in the coronet of rank, or stains the ermine of justice, or skulks in the cleft of the mitre, or is wrapped up in the senatorial robe, or cankers the green wreath of genius ; when acts of political corruption or public immorality are mingled with individual, domestic, or social vices, courting imitation or applause, and offering violence to the feelings and principles of the community, it becomes the duty of the patriot and the moralist to hold up to public shame the enemies of public virtue. Such a patriot and moralist was Junius. The flash of his mental eye scathed as with a lightning-stroke the minions of corruption, and men paused in their career of political mischief in order to avoid the fate of his victims. Envenomed with wit and winged with sarcasm, his shafts carried dismay into the ranks of his adversaries, and they struck deeper into their prey in proportion to the polish with which they had been elaborated ; and when he failed to annoy and dislodge his antagonist by the light troops of his wit and ridicule, he brought up in reserve the heavy artillery of a powerful and commanding eloquence. In thus discharging the duties of a public censor, and in defending, at the risk of his life, the laws and constitution of his country, we may admire the courage of Junius, and even proffer to him our gratitude, though we disown his political principles and disapprove of his conduct. The good done by Junius has lived after him ; let the evil be interred with his bones. But though the character of Junius, while he himself remains in the shade, may be pure and noble, it may assume a different aspect when he is identified. Were Lord Chatham or Lord Sackville, or Burke or Sir Philip Francis to stand forth as Junius, his morality would disappear, and his patriotism sink into disaffection and disloyalty ; and were either Barré or Macleane to be honoured with his laurels, we must brand them as traitors to the cause which they advocated, and as men who bartered their obligations to the community for a mess of pottage.

"It is always instructive, and now more than ever, to *beware of patriots*, to scrutinize the pretensions of popular leaders, and to estimate the value of their labours. Junius was a very moderate reformer, liberal in his political views, but hostile to innovation. His object was to defend constitutional rights, and not to create them. It was ' *the unimpaired hereditary freehold*' which he strove to bequeath to posterity. It was the 'liberty of the press—the palladium of all the civil, political, and religious rights of Englishmen,' and the right of juries to return a general verdict, for which he combated. Had he lived in the present day he would neither have been a Repealer nor a Confederate nor a Chartist. He would have hesitated even to extend the suffrage till the people were fit to exercise it, for he declared that both liberty and property would be precarious till the people had acquired *sense* and spirit to defend them. Education and religious knowledge must precede the extension of political privileges. No person is entitled to a political right till he has learned how to use it ; no man is qualified for a trust till he knows how to fulfil it. The rights of the subject are not the rights of an individual, but the rights of the community ; and he who either prostitutes or sells such a birthright, dishonours and robs every member of the community to whom the same inheritance has been bequeathed."

The following request and response are in letters without dates, but must have been written about this time. Miss Edgeworth writes :—

" I am writing a philosophical tale upon the dangers and follies, tragic and comic, of ' Taking for granted.' I wish you would help me to a few good instances either in science or common life. I asked Sir H. Davy, and he gave me one on demand. People having at first *taken for granted* that stones, commonly called thunderstones or moon-stones, *could* not have fallen from the moon or planets, prevented all inquiry or reasoning, and made others take for granted they must be thrown from volcanoes. I am sure you could furnish me with better instances."

Dr. Brewster gave a ready response :—

" I fear my instances of ' taking for granted' may have too much prose or pedantry in them for your use, but if they suit you I could easily give you more, as almost

all the grave blunders which mark the history of science
are referable to that source of error. The most remarkable example that has ever fallen in my way is that of
Sir Isaac Newton in his celebrated analysis of the Prismatic Spectrum. He took it for granted that at equal
refractions the length of the spectrum was the same for
all substances of which his prism could be made ; or, in
other words, that all bodies had the same dispersive
power. So thoroughly had he taken this for granted,
that his eye seems to have been blind upon this subject
alone ; for though he used spectra formed by water,
and by different kinds of glass, yet he never saw, what
is easily seen, that they are very different in magnitude
at equal angles of refraction. Hence he missed the
great discovery of the achromatic telescope, which he
must have invented instantly had he not indulged in
'taking for granted.' Nay, he was so convinced that
there was no difference in the spectra, that he actually
prevented others from making the very experiments
which every person presumed he must have made.

" It is now curious to observe, that while taking for
granted deprived Sir Isaac of the honour of inventing
the achromatic telescope, the very same invention was
afterwards made in consequence of Euler taking for
granted that the human eye was an achromatic instrument, which it is not. Conceiving that all the works of
the Almighty must be perfect, he conceived that the
errors of colour must be corrected in the human eye,
forgetting that the supreme wisdom might be evinced
in making that organ perfect without any such correction. He therefore set about computing the curvature of lenses of glass and water for making telescopes, and out of the experiments and inquiries which

sprang from this investigation of Euler arose the achromatic telescope, one of the finest inventions of modern times.

" It was then discovered, and proved beyond a doubt by Dr. Blair, that the opposite spectra formed by two kinds of glass, or two fluids, could not correct one another, as in each spectrum the coloured spaces had different proportions. The result was that there was a secondary spectrum left which affected all achromatic telescopes. The remedy for this Dr. Blair discovered, and he produced what he called a *planatic* telescope, in which all colour was perfectly corrected by three media or more. Notwithstanding all this, Dr. Wollaston took it for granted that what he could not detect with his eye could not exist ; and, in a paper on the spectrum, broadly asserts that in all the spectra which he could form, the colours had the same proportion in similar positions of the prism. This was a great mistake ; the effect, which he did not see, required to be looked for carefully, and, in some cases, to be magnified; and there is no doubt that Dr. Blair's doctrine is the right one. England used to supply the whole world with achromatic telescopes, but from the want of glass it became difficult to make them of a large size. It was literally taken for granted that flint-glass could not be made pure and free of veins in large pieces, and nobody ever made any attempt to manufacture it. A poor Swiss peasant imagined that he could succeed, and after many trials he was perfectly successful. He taught the art to Frauenhofer of Munich, who has made the most magnificent telescopes that were ever seen, and deprived England not only of her trade in this article, but of all her practical glory.

" You see that I cannot keep myself out of a dissertation, so that I shall not make any more such demands upon your patience, unless you tell me that such examples as the above will suit you."

Another source of great interest arose from the following lively request of Miss Edgeworth's, dated Oct. 9th, 1823 :—

" . . . I hope Mrs. Brewster's health continues to improve, and that she is not now confined to her sofa.    The half hour I spent with her was very short and very sweet.    I hope the boys go out sometimes without their great-coats !    I am much tempted to make a bold request of you, and to give you what I know you hate above all things—a great deal of TROUBLE. I am writing a sequel to my father's ' Harry and Lucy,' which I am naturally desirous should be worthy of his beginning.    But this is a hard battle to me, as I have not, what I once heard a gentleman boast he possessed, ' *a leetle scientific knowledge that came naturally.*'    Whatever I have, came very unnaturally, and with hard labour, and most of it from your *Encyclopædia*, and others.    My wish would be to send you some parts of ' Harry and Lucy' to look over before going to press. But from this I am deterred by the hatred which I feel myself to looking over other people's MSS. and correcting them.    I know, moreover, from your own dear bosom friends (who by the bye are always the people who tell one's snug faults) that you are one of the most indolent philosophers extant—a bold word—and that you never can bring yourself to finish your own undertakings for the press, even till the last gasp, though all the printers' devils are waiting for and urging you !

" Then how much the less could I expect that you should correct for me, who am neither devil quite, nor angel quite, nor anything at all to you?

" Then, supposing your good-nature, chivalry, Quixotism, or some of what becomes a Scottish gentleman-philosopher, *should* work and egg you up to accept this troublesome task, which all the while would go against your natural stomach, there remains the difficulty of transmitting the pages of the MS. to you. Not a soul or body near but has too much conscience to be obliging, as they used to be, in franking large packets, before the Reformation and the Commissioner came to Ireland. So I give it up, and only write this to divert you some fine day when you have grown too pale over the midnight lamp."

The plan was not given up, however; officials proved obliging, the "Scottish gentleman-philosopher" was characteristically accessible, and all the MSS. of the "Sequel to Harry and Lucy" were subjected, not only to Dr. Brewster's scientific criticism, but were read aloud to his four boys, in order to judge of the effect of the mixture of science and story on juvenile minds. The intense interest it excited was a true presage of the popularity which, when published, it continued so deservedly to meet with from all intelligent youthful readers. The philosopher threw himself into this occupation with all his heart, as is evident from the letters of judicious advice, criticism, and encouragement which accompanied each returned packet. I give some interesting extracts from these :—

" *Feb.* 7, 1824.

" I have safely received your different packets. I

now return Nos. 4, 5, and 6, which have afforded much
pleasure to my young people, and which I am sure
they thoroughly understand. I do not think they are
capable of improvement. I would only suggest the
introduction of the Chinese fish or serpents (as they
are called) which are made of thin ivory, as a striking
illustration of the effects of moisture. . . . In two days
you will receive Nos. 7-10. After succeeding with the
steam-engine, I expect that you will soon be found
trespassing upon my manor among telescopes and microscopes and double refractors. If you continue to let
the game pass through my hands, I shall be gentle with
you before the justices, and have no doubt that they
will reverse the order of their judgment by being less
severe upon the last than upon the first offence."

> " 10 COATES CRESCENT, EDINR.,
> *March* 19, 1824.

" MY DEAR MISS EDGEWORTH,—I return your packet,
and have been much gratified with the very perspicuous
account which you have given of the phenomena of
mother-of-pearl. The only remark that occurs to me
is, that the reader might suppose that there are no other
colours in mother-of-pearl but the superficial or communicable ones, while in reality there are others which
are produced in the interior, and are therefore not communicable to wax.

" There is also a very extraordinary fact respecting
these communicable colours in mother-of-pearl which
deserves to be mentioned. One set of these colours is
produced by the right side of the grooves, and another
set by the left side of the grooves, and *both* of them are
distinctly seen when the mother-of-pearl is *polished* :
but when the polish is removed by rough grinding, one

of the sets *invariably disappears.* The rough grinding, therefore, destroys the effect of one side of the grooves without affecting the other, a result which I have never been able to explain satisfactorily.

" Mrs. B. and I will read with much pleasure the work you mention, and I am most anxious to see Ireland with my own eyes. I shall certainly make a great exertion to obtain this gratification this summer, but the exertion is great to those who, like me, are entangled in business, in country concerns, with children and printers' devils, and a lawsuit to boot."

" 10 COATES CRESCENT, EDINBURGH,
*April* 2, 1824.

" . . . I am almost afraid to put down in writing another observation, and yet it would be a weakness to omit it. The passage you quote from the Scotch novel relative to Mr. Watt is quite overstrained, and, in every respect, utterly incorrect. No man ever admired Mr. Watt more than I did, and I was peculiarly fortunate in corresponding with him during the last years of his life, and enjoying a good deal of his society, both here and at his own house. I admired him, however, for what he did, and not for what he never thought of doing ; and I confess that I have been guilty, under the influence of personal attachment, of writing of his labours in very exaggerated strains. Mr. Watt's great invention was to improve the steam-engine by the use of a separate condenser; but great as this was, I am convinced that steam-boats and steam-engines would have been in the same state of perfection at this moment had Mr. Watt never lived. I do not mean to say that any person would have improved the low-pressure steam-engine by the invention of a separate condenser; but I maintain that the high-pressure engine, where there is no condensation at all, and which Mr. Watt and all his friends (including myself) reviled as of inferior utility, is in every respect superior to the low-pressure engine, and would have accomplished all the great operations of modern times, even if the low-pressure engine and a separate condenser had never been heard of.

" Nay, I am disposed to think that the obstinate adherence of Mr. Watt and his friends to the low-pressure engine, long after accurate experiments made and recorded in Cornwall, had demonstrated the superiority of high-pressure ones, has done much to retard the progress of invention respecting this engine. The same obstinacy is at this moment opposing the invention of Mr. Perkins, which is a step as far above the high-pressure engine as that was above the low-pressure one, even if Mr. Perkins realizes only one-third of the power which he expects.

"The author of Waverley has, moreover, forgotten the practical excellence of Boulton's steam-engine, which drained mines and coal-pits as successfully as Mr. Watt's, from which it differed chiefly in having a greater appetite for fuel.

" When the poet or the orator is called upon to declaim upon any great national invention, his art requires that it should be associated with one great name. The imagination can as little tolerate the subdivision of praise, as it does that of labour; and hence, whenever the objects of science or art come within its domains, all individuality is lost either in the brilliancy or obscurity of its creations.

" You would oblige me much by giving me the information you mention on double stars, in which I take a very great interest. I suppose the discoveries are those of Mr. Herschel, who has been long occupied with this curious subject."

Miss Edgeworth seems somewhat to have misunderstood Dr. Brewster's plain speaking, as was often the case with his statements, which were generally made with no other thought than the subject immediately before him. He writes, in answer to a letter of hers :—" I owe you one grudge, however, for supposing that I would not be pleased with the meeting in honour of Mr. Watt. I am sure that I expressed myself hastily on that subject, otherwise you could not have misunderstood me ; but I am glad to be able to say that I had attended the Edinburgh meeting on the very day preceding that on which I was favoured with your letter." If not able to see eye to eye with many of Mr. Watt's friends on the subject of the steam-engine, it is pleasant to see the full justice which Brewster did to Watt long afterwards upon the priority of his discovery of the Composition of Water. The biographer of Watt thus writes :—" As an instance of the change which was wrought by the force of truth on the convictions of others equally distinguished, we may mention a most eminent philosopher, who having, at a former period, on the imperfect information then open to him, been disposed to support the claims of Cavendish, on fully studying the fresh evidence which the correspondence of Mr. Watt first made public, unhesitatingly professed his entire conversion ; and in one of those eloquent essays by which he has so often adorned the progress of scientific discovery, publicly announced, as the conclusion at which he had arrived, that the argument for Mr. Watt's priority ' had now been placed on a sound and impregnable basis.' That the name of Sir David Brewster should be known throughout the whole civilized world by the most brilliant discoveries in the most beautiful of sciences, can scarcely be deemed more honourable to him as a man, than the perfect candour which he thus displayed; and such unreserved testimony, spontaneously borne under such circumstances, by such an authority, has evidently a most conclusive bearing on the question in regard to which it was delivered."

Another strongly expressed opinion meets us in this letter, which he never receded from, as the same sentiments are as decidedly stated in his *Memoirs of Sir Isaac Newton*, published more than thirty years afterwards :—

"*April* 26, 1824.

" . . . I would strongly recommend the omission of the passages about Lord Bacon in pages 703, 704, even if you should not agree with me in the opinion which I am about to state. The opinion so prevalent during the last thirty years, that Lord Bacon introduced the art of experimental inquiry on physical subjects, and that he devised and published a method of discovering scientific truth, called the method of induction, appears to me to be without foundation, and perfectly inconsistent with the history of science. This heresy, which I consider as most injurious to the progress of scientific inquiry, seems to have been first propagated by D'Alembert, and afterwards fostered in our University by Mr. Stewart and Mr. Playfair, three men of great talent, but not one of whom ever made a single discovery in physics.

"It is an undoubted fact that Kepler, Galileo, and Huygens were as well acquainted with the method of conducting scientific inquiries as any philosophers that have flourished since the time of Bacon ; Dr. Gilbert of Colchester, before Bacon's time, gave the most perfect specimens of the method of investigating truth by experiment and observation; and he speaks in the strongest language of the absolute inutility and folly of all other methods. Leonardo da Vinci too speaks in the strongest terms of the omnipotence of experiment and facts in all philosophical pursuits. It has been said, however, by the admirers of Bacon, that though a few philosophers knew the secret of making advances in science, yet the great body were ignorant of it, and that Paracelsus, Van Helmont, and many others, were guided in their inquiries by very inferior methods. In answer to this argument, it is sufficient to say that in the present day there are numbers of philosophers who are quite ignorant of the proper method of conducting physical inquiries, and who follow their own whims and fancies as much as Paracelsus and Van Helmont.

"If Bacon introduced any new method into science, it seems very strange that his contemporaries never thanked him for it, nor seemed to be aware of it. Sir Isaac Newton, who is invariably said by some modern writers to have made all his discoveries by following Bacon's rules, never once mentions either Bacon or his method, and the good Mr. Boyle is equally silent on the subject. If Bacon had never lived and never written, science would have been just where it is at this moment.

"It seems quite clear that Bacon, who knew nothing either of Mathematics or Physics, conceived the ambitious design of establishing a general method of scientific inquiry. This method, which he has explained at great length, is neither more nor less than a crusade against Aristotle, with the words *experiment* and *observation* emblazoned on his banner. This hue and cry about experiment was so far good, that it was in a good cause. But the cry was old; all men of talent had obeyed it, and those who did not, namely, the charlatans of the day, were neither willing nor able to renounce their speculations and extravagancies. These empirics had in former times some hold on the public mind, which was then illiterate and ill-informed on all points, whereas the same class of persons, who are as numerous as ever, are kept in the background from the wide diffusion of sound knowledge and the prevalence of sober opinions among all orders.

" The method given by Bacon is, independent of all this, quite useless, and in point of fact has never been used in any successful inquiry. A collection of facts, however skilfully they may be conjured with, can never yield general laws unless they contain that master-fact in which the discovery resides, or upon which the law mainly depends. It is often some hidden relation, some deep-seated affinity, which is required to complete, or rather to constitute, a great discovery; and this relation is often discovered among the wildest conceptions and fancies after they have been sobered down by the application of experiment and observation. The extravagant speculations which often precede and lead to discovery differ in no respect from the creations of a rich poetical fancy. Wild and unsubstantial in them-

selves, they pass over the mind like a shadow, and it
is only when they are clothed in the imagery of external nature, and invested with the realities of human
feeling, that they begin to exercise their power over the
heart.

" In short, it is just as true that Scott and Byron composed their works under the tuition of Horace's *Art of
Poetry*, as it is that Newton made his discoveries by
following the method of Bacon.    Horace, too, was a
poet, and capable of laying down some technicalities
for the advantage of future bards ; but Bacon was no
natural philosopher, and has even demonstrated the
utter inanity of his method by the ridiculous results to
which he was led in applying it to the subject of heat.
The application of Horace's maxim of *nonum in an-
num premere* would not more effectually extinguish all
modern poetry than the application of Bacon's method
would extinguish all modern science."

The following short extracts from some of the other
letters may be given, as showing the different topics of
interest which were crossing Brewster's path at the
time ; the mutual friendship felt for " Sir Walter" showing itself by frequent allusions throughout the correspondence.    There appears to have been a frequent
interchange of plants and minerals between Edgeworthstown and Coates Crescent :—

" I have forgotten to thank you for the American
moss, which derives double interest from its description
in *Harry and Lucy*.    I have enclosed some specimens
of Tabasheer, a substance of extreme rarity and great
interest as a sort of mineral product of vegetation," and
in the next letter—" I do not remember if I mentioned

I

to you when I sent you some Tabasheer, the singular phenomena that relate to the distribution of silex in the cuticle of several plants, and which gives them the power of polishing wood and even brass. I enclose a brief printed notice of what I observed, which is given in Dr. Greville's *Flora Edinensis.*" The following is what was enclosed :—

"*Equisetum Hyemale—Rough Horsetail.*—This species contains more silex beneath its delicate epidermis than any other, and is consequently most employed in polishing hard wood, ivory, and even brass. The silex is so abundant, that the vegetable matter may be destroyed, and the form retained, as was effected by Mr. Sivright.

"My friend Dr. Brewster has obligingly permitted me to consult an unpublished paper, written by him on this subject. On subjecting a portion of the cuticle to the analysis of polarized light under a high magnifying power, Dr. Brewster detected a beautiful arrangement of the siliceous particles, which are distributed in two lines parallel to the axis of the stem, and extending over the whole surface. The greater number of the particles form simple straight lines, but the rest are ' grouped into oval forms, connected together like the jewels of a necklace, by a chain of particles forming a sort of curvilineal quadrangle ; these rows of oval combinations being arranged in pairs.' Many of those particles which form the straight lines do not exceed the 500th part of an inch in diameter. Dr. Brewster also observed the remarkable fact, that each particle has a regular axis of double refraction. In the straw and chaff of wheat, barley, oats, and rye, he noticed analogous phenomena, but the particles were arranged in a different manner, and ' displayed figures of singular beauty.' From these data, the Doctor concludes 'that the crystalline portions of silex, and other earths which are found in vegetable films, are not foreign substances of accidental occurrence, but are integral parts of the plant itself, and probably perform some important function in the processes of vegetable life.'"

"*Jan.* 8, 1827.

" Napoleon is now finished, excepting the Appendix. It has annoyed Sir Walter more than any of his other productions. He said lately to a friend that he did not know whether he would finish Napoleon or Napoleon finish him. . . . Can you tell me if it is the opinion in Ireland that the salmon fry which go down to the sea return in the shape of grilse ; or in the shape of sea trout, finnocks, or whitlings ?"

"*March* 24, 1829.

". . . I have such a love for Ireland that I would fain congratulate you, if I could, on your prospects. You are under the influence of Paradise gas, which, I fear, will neither fill your bread-baskets nor cover your epidermis, nor change your masters. When the imagination has had its triumphs, your cottagers will, I fear, discover that in the division of the once forbidden fruit, the rich have seized the kernel and given them the shell. How happy shall I be to confess to you some years hence the delusion under which I write !

" Sir Walter Scott is just about to make a fortune by printing his novels at a cheap rate for general circulation. Why should *you* not make ten thousand pounds by doing the same ?"

## CHAPTER IX.

### NOTES OF LIFE FROM 1824 TO 1830.

THERE is no flock, however watched and tended,
  But one dead lamb is there !
There is no fireside, howsoe'er defended,
  But has one vacant chair !

Let us be patient ! These severe afflictions
  Not from the ground arise,
But oftentimes celestial benedictions
  Assume this dark disguise.
                              LONGFELLOW.

IN 1825, Brewster was made a Corresponding Member
of the French Institute, and from this time honours
crowded in so rapidly upon him, that except any of special interest it would be tedious to enumerate them
in their order and succession.   Suffice it to say, that
the large book in which the letters, diplomas, burgess
tickets, announcements of medals, etc., are collected,
is a remarkable one for size and value.   The large
towns of Switzerland, France, Germany, Holland, Italy,
Russia, Belgium, Portugal, Austria, Sweden, and Nor-
way, South Africa, Antigua, the various States of
America, besides the towns and Universities of England, Scotland, and Ireland, all contributed their quota
of honours to this man of research and industry.   A
cape received his name in the Arctic regions, a river
in the Antarctic, and a new plant discovered by Dr.

Muellin in Australia was named *Cassia Brewsteri.* He received, besides the Copley, Rumford, and Royal Medals, two Keith Medals from the Royal Society of Edinburgh, two from the French Institute, one from Denmark, one from the Société Française de Photographie, and various others ; of some of the most valuable of these, duplicates were sent to him, one of gold, which he turned into plate, and a facsimile of frosted silver,—all being preserved as heirlooms.

Many men of eminence in all departments of science and literature naturally gathered around Brewster during his residence in Edinburgh, and many followed him to the country with letters of introduction, while most of them became contributors to some of his editorial works. One of these, M. Audubon, the naturalist, has left a pleasing tribute to the home *bienséances* of the philosopher, whose wife he records as " a charming woman, whose manners put me entirely at ease." Another literary acquaintance of this period, Dr. Hibbert, the author of a book *On Apparitions*, which was well known in its day, visited Dr. Brewster at Allerly. A trifling but amusing incident which occurred on this occasion he was fond of relating to the latest years of his life. The learned guest had retired to his chamber, while the learned host pursued his researches in his study, when a loud outcry was heard in the house, which was wrapped in the silence of the small hours. Dr. Brewster rushed into the lobby, where a strange scene presented itself. The author of the book *On Apparitions*, forgetful of its scientific reasonings, was standing at the door of his room staring wildly around and ejaculating " that there was a very curious light somewhere !" while, at another door, two small boys in

their night-dresses, roused at first in affright, were in
convulsed laughter, for the ghostly light arose from
the thick tassel of the author's night-cap, which had
been set on fire before quenching the midnight lamp,
and which was now smouldering on the top of his
head.

In 1827 the long talked-of visit to Edgeworthstown
was at last made out, and Brewster also visited several
members of the family in other parts of Ireland, but he
spent the greater portion of his time in the home of the
Edgeworths, so celebrated as a centre of literature, of
persevering work, and of happy home life.    He writes
as follows to his wife :—

" EDGEWORTHSTOWN, *July* 17, 1827.

" We stayed all Friday night at the Observatory of
Armagh, at Dr. Romney Robinson's, the astronomer, a
person of most extraordinary talents.  My visit here
has been a very agreeable and delightful one.  A more
extraordinary family for talents, mutual affection, and
everything that can interest, I could not have conceived.
You must come to Ireland to see what will never be seen
again.  We sit down to dinner, a family party of sixteen,
and a merrier one you have never seen.   The country
about Edgeworthstown is not at all to be compared
with the county of Antrim, but the place itself is
fine, and the lawn extensive and grand.  Miss Edgeworth is to send with me some plants for you, and
I hope to get what is very rare, a white rhododen-
dron from Ballydrain, from whence she is also to
get one.   I have seen a good deal of Ireland during
the short time I have been here.   It is in every
way a most extraordinary country.  I like the people
amazingly, and have no more fear of travelling at night

than if I were in Roxburghshire. We have been busy
here all day with meteorological observations, which
have been made principally by Miss Fanny, a most
elegant and accomplished person. I hope Charles has
not forgotten them. Yesterday Mr. William Edgeworth, M. Davidoff, Mr. Collyar, and I drove to Loch
Gorma, a lake about twelve miles distant, so that with
that journey, and our visit to the school, etc., here, I have
scarcely leisure to scribble this epistle. Miss Edgeworth is to send some curiosities to Harry."

In this year the family residence was changed entirely
to Allerly, which possessed many advantages of health,
comfort, and leisure. The society of Roxburghshire at
that time presented a most extraordinary circle of
literature. Abbotsford itself was a sun and centre from
whence the great enchanter drew to himself, with the
cords of his mighty genial genius, those eminent in
kindred pursuits from every part of the world, not to
speak of the inferior crowds of "beasts with sketch-
books," as his lively daughter, Anne Scott, designated
the followers of the great lion. The country-houses
around Abbotsford were also peopled with authors and
wits. Chiefswood, a lovely cottage *ornée* within a short
walk of Abbotsford, up a wooded glen, was tenanted first
by Captain Hamilton, author of *Cyril Thornton*, to whom
Mrs. Hemans, the poetess, paid a long visit, and afterwards by Lockhart, the son-in-law of Scott, and himself celebrated as a critic and author. Huntly Burn
was the residence of Sir Adam Ferguson, the chosen
friend and companion of Sir Walter Scott, whose witty
stories and *bon mots* mixed well with the other literary
ingredients. At Maxpoffle, some miles further, G. P. R.

James, the well-known historical novelist, pitched his
tent, and Mr. Poulett Scrope, author of *Days of Deer-
stalking* and *Nights of Salmon-fishing*, who was also an
excellent amateur painter, resided at the Pavilion for
some years ; and if we add to the list the philosopher of
Allerly, with his science and occasional *savans*, it would
be difficult to find a country circle in which so many
men of mark were included.

There was much sociability in the neighbourhood
of Melrose.   Some of the families above mentioned,
among whom were Sir David and Lady Brewster and
other neighbours, formed a club which they called
" The Barley Broth Club," the rules of which laid its
members under the pleasant necessity of dining alternately at each other's houses every Saturday, and these
happy genial reunions are recalled with regretful pleasure by the few survivors.[1]

The friendship and intercourse between the families
of Abbotsford and Allerly was intimate and frequent.
My own memory goes back as in a dream to that time
—the men of mail and the armoury, the dogs of flesh
and stone, the galleried study, the favourite invalid
grandson, " Hugh Little John," and kind words and
caresses from the Great Wizard himself.   Nowhere was
the throb of sorrow that pervaded the nation in 1832
more felt than at Allerly, when, on the 21st of September, the honoured head lay down in its melancholy
last slumber on that " beautiful day, so warm that every
window was wide open, and so perfectly still that the
sound of all others most delicious to his ear, the gentle
ripple of the Tweed over its pebbles, was distinctly

---

[1] This club is mentioned in a letter from Dr. Clarkson, one of its members, to Sir J. Y. Simpson, Bart.

audible as we knelt around the bed, and his eldest son
kissed and closed his eyes."[1]

Before that time, however, the cloud of sorrow had
gathered over Allerly itself deep and dark.   There was
no mill-stream near the house, as at the dreaded Allars,
but the full rippling Tweed rolled its waters in sufficient
proximity to be tempting for boyish sports.   A strange
fear of drowning had pervaded Dr. Brewster's life.   He
always believed that he himself was to perish in that
way, a fear which strangely enough was discovered to
haunt the mind of more than one of his descendants,
even when too youthful to be prepossessed by any
knowledge of others having felt the same.   But the
father's fears and caution seem to have slumbered for a
while, and the young Brewsters were frequently allowed
to bathe in the river with companions.   The second
son, Charles, was a boy of fifteen, of the highest promise for steadiness and talent,—one of those beings
so universally beloved and so perfect in home life that
it is commonly, though wrongly, said, " They are too
good to live," as if the grace of God could not fit for
life as well as death.   One summer's afternoon, in 1828,
the sudden, fearful summons came to the agonized
parents from the neighbour's house to which the unconscious boy had been conveyed from the fatal river;
but, alas! life was proved to be extinct before they
could reach the house, and the young fair form of their
cherished "Charlie," that went forth at noon in health
and strength and beauty, was brought back ere evening
a lifeless corpse to the desolated home.   It was Dr.
Brewster's first great sorrow, and it was his most overwhelming.   A year afterwards Sir Walter Scott wrote

1 See Lockhart's *Life of Scott.*

to Miss Edgeworth :—" Dr. and Mrs. Brewster are rather getting over their heavy loss, but it is still too visible on their brows, and that broad river lying daily before them is a sad remembrancer. I saw a brother of yours on a visit to Allerly."[1]

The slightest mention of death by drowning was an agony. For many years neither parent ever mentioned the " once familiar word," and even when long years of time allowed of the occasional utterance of the beloved name, it was never without the peculiar thrill and almost sob of an ever-living grief.

Some years after, his son David was very nearly lost in Duddingston Loch, but the narrow escape was carefully kept from his father's knowledge. About thirty years afterwards the following incident occurred, which his daughter-in-law, Mrs. Macpherson, thus relates :— " One evening I was showing him an album which had been sent for his autograph. When we came to a picture of Duddingston Loch, I said, ' Oh, that is where David was so nearly drowned.' He had never heard of the incident, and I never saw him so overpowered as when I told him the story. I suggested that perhaps he had forgotten it. ' No, indeed,' was the answer, 'it would have made an awful impression upon me after his brother's death.'"

Of the real effects of this bereavement upon Brewster's mind and spirit we know little. Mr. Veitch hastened to see and sympathize with his old friend soon after the heavy trial. He was deeply and thankfully impressed with the apparent deepening of religious thought in the bereaved father, who had been reading Baxter's *Dying Thoughts* with extreme interest, and gave a copy

[1] Lockhart's *Life of Scott.*

of it to his friend. After this Mr. Veitch frequently
visited Allerly on less melancholy occasions. I remember well the fine old man of science, with his racy
Scotch and massive figure, while the two grey-headed
men enjoyed together much of the pleasant intercourse of the old, old days. Correspondence was still
carried on at intervals, but Mr. Veitch was a much
older man for his years than the ten years' of seniority
to his friend warranted; failing health began to set
in, and symptoms of bodily decline interfered with the
old activity. The mind, however, and the fingers retained their pristine vigour. The last work that he
finished, just in time to lie down immediately afterwards on his dying-bed, was two thermometers, finished
with all his wonted accuracy and delicacy of execution.
Notwithstanding the bodily prostration which rapidly
ensued, his mind retained its full powers until two days
before the end, and then it wandered into the belief,
full of a solemn and joyful truth, that he was on a long
journey, and that he was going home. The workings of his mind before that slight final wandering
seem to have been most characteristic; from the first
seizure he felt that he had no more to do with the
world, and devoted himself to fresh preparation for the
great change, which, however, had never been forgotten
in the midst of science, mechanics, and busy life. "He
spoke repeatedly of the mystery of man's being, and
the close alliance, yet clear distinction, between mind
and body, testing the continued soundness of intellect
by his ability to go through a process of calculation.
His concern, however, was chiefly about the things of
his peace; he expressed his deep sense of sinfulness and
his trust in redemption through the blood of Christ. He

frequently requested the Scriptures to be read, especially the Psalms and passages from the Gospel of John, and also to have prayer offered in which he might join; he seemed himself to be privately much engaged in commending himself to the mercy of the Saviour. Of his approaching end he spoke with calmness and solemnity, and, if not with assurance, yet with good hope through grace. At length, on the morning of the 10th day of June 1838, his strength completely failed, and he quietly departed to his rest in the sixty-eighth year of his age. Though unable to speak, he evidently retained his consciousness till nearly the last breath. Of keen temperament, his language might be at times strong, but he ever honestly expressed his convictions without regard to fear or favour. He was warmly attached to the Church of Scotland, of which he was an elder; he venerated the memory of her martyrs, and amidst tokens of change was zealous for the maintenance of her purity of doctrine and worship. The writings of the old divines were his favourite study, especially on the Lord's day. He regularly maintained family worship, and through life it had been his desire and endeavour that, whatever others did, he and his house should serve the Lord."[1]

[1] From Notes by Rev. Dr. Veitch.

# CHAPTER X.

## NOTES OF LIFE FROM 1830 TO 1836.

A THOUSAND glorious actions that might claim
Triumphant laurels and immortal fame,
Confused in crowds of glorious actions lie,
And troops of heroes undistinguished die.
ADDISON.

But what on earth can long abide in state,
Or who can him assure of happy day ?
Sith morning fair may bring foul evening late,
And least mishap the most bliss alter may ?
For thousand perils lie in close await
About us daily, to work our decay,
That none, except the God of heaven him guide,
May them avoid, or remedy provide.
SPENSER.

THE next few years of the philosopher's life present
some changes and vicissitudes, and many troubles, but
were pervaded as usual with the element of assiduous
work.  Chiefly through his energy and unwearied per-
severance a memorable prophecy uttered by Lord Bacon,
that for the better development of intelligence and
learning there would be established " circuits or visits
to divers principal cities of the kingdom," began to
have its fulfilment.  The decline of science, and the
small encouragement given to scientific men in England, had excited much attention and discussion.
Previous to 1826 Sir John Leslie and Professor Playfair had expressed strong opinions ; between that year
and 1831, Sir Humphry Davy, Lord Brougham, Sir
John Herschel, Mr. Babbage, Dr. Daubeny, and other

men of science, had successively written on the subject, while Mr. Douglas of Cavers, in his *Prospects of Britain*, devotes an admirable chapter to the "Decline of Science and the means of its Revival." Brewster proposed to the world—in a review in the *Quarterly* of Mr. Babbage's work, *Reflections on the Decline of Science in England, and on some of its Causes*—a plan for the remedy of this evil, which he described in the following language:—"AN ASSOCIATION *of our nobility, clergy, gentry, and philosophers*, can alone draw the attention of the Sovereign and the nation to this blot upon its fame. Our aristocracy will not decline to resume their proud station as the patrons of genius; and our Boyles and Cavendishes, and Montagues and Howards, will not renounce their place in the scientific annals of England. The prelates of our National Church will not refuse to promote that knowledge which is the foundation of pure religion, and those noble inquiries which elevate the mind and prepare it for its immortal destination! If this effort fail, we must wait for the revival of better feelings, and deplore our national misfortune in the language of the wise man: 'I returned, and saw under the sun, that there is neither yet bread to the wise, nor yet riches to men of understanding, nor yet favour to men of skill.'" In the course of a few succeeding months the plan of the "British Association for the Advancement of Science" met with general acceptance, and was soon thoroughly matured, the first meeting taking place at York, in September 1831. The arrangements were much like those of succeeding years. It lasted for a week, during which morning meetings were held, at which scientific papers were read and oral communications discussed, upon the branches of science comprised within the different sections; more popular lectures and exhibitions of interesting objects being generally reserved for the evening. This meeting was a decided success. Fresh vigour and interest in science, mutual sympathy and quickened intelligence certainly arose from its animated discussions and its free exchanges of thought and discovery. The families in the neighbourhood of York took a great interest in the scientific assembly, and the Archiepiscopal Palace at Bishopthorpe was thrown open to its members.

Dr. Brewster thus writes to his wife :—

"BISHOPTHORPE, *Sept.* 30, 1831.

" I sit down at one o'clock in the morning to write you a legible letter, which I fear the one I wrote you yesterday could scarcely be called. I came here to-day to dinner, and was most kindly received by the Archbishop, who made me feel at once that I was at home. He and the whole of the party here returned to York to hear Mr. Scoresby's lecture on his new magnetical discoveries. The assemblage of beauty, fashion, and philosophy was really splendid, and after eleven o'clock we returned to the Palace. To-morrow we again go to York after breakfast, and after spending the whole day in the arrangement for a *General British Association for the Advancement of Science,* and in hearing many scientific papers, we return to dinner as we have done to-day.

The success of the meeting has infinitely surpassed all our most sanguine expectations. No fewer than 325 members, have enrolled their names, and a zeal for science has been excited which will not soon subside. The next meeting is to be held at Oxford, in June, at the time of the commemoration, and in the Radcliffe Library

or the Theatre. . . . The Archbishop, after reading a
letter from the Archbishop of Canterbury, remarked to
me that it was not yet known how the Lords were to
act.[1] He added that the Archbishop of Canterbury
had not made up his own mind, and that he held his
proxy, so that it is clear that the Archbishops have not
decided against it. Lord Morpeth's letter stated that
several of the Lords were not to vote at all, and it
seemed to be the opinion that the Reform Bill would
be carried by the neutrality of those who might be
expected to oppose it. What a charming and princely
spot this is, as much from its ancient and splendid
apartments as from the richness and variety of the
grounds! The Archbishop has invited fifty or sixty of
the philosophers to dine here to-morrow, among whom
are Sir T. Brisbane, Thos. Allan, and the rest of our
Scotch party. . . . Mr. Vernon Harcourt, the soul of
our meeting, and one of the most amiable and learned
of men, is the eldest son of the Archbishop."

The York meeting did not pass off, however, without
a slight cloud, which threatened to injure the main
object of the Association. Lord Milton, the President,
in his opening speech, to the surprise of his audience,
made himself understood as objecting " to all direct
encouragement of science by the State," characterizing
such a mode of advancing it " as un-English." Mr.
Vernon Harcourt, however, the Vice-President, replied
in the following corrective words :—

" I should undoubtedly be very sorry to see any system of encouragement adopted by which the men of
science in England should become servile pensioners of

[1] This was the time of the first Reform Bill.

the Ministry; and no less sorry am I to see them under
the present system, when exerting the rarest intellectual faculties in the scientific service of the State,
chained down in a needy dependence on a too penu-
rious Government. . . . As things stand at present, the
deeper, drier, and more exalted a man's studies are, the
drier, lower, and more sparing must be his diet. . . . I
cannot see any reason why, with proper precaution,
men of science should not be helped to study for the
public good, as well as statesmen to act for it. It cannot be wondered at that our philosophers should be
unwilling to hear it proclaimed *ex cathedrâ*, from the
midst of themselves, that there is something illegiti-
mate in the direct encouragement of science, though
they are ready enough to own that there is something
in it very '*un-English*.'"

It is often said that a meeting of the British Association is merely a pleasant conversational week, which
when over leaves no practical effect on society. Several
purely scientific objects have, however, been carried out.
Useful and encouraging reports on the state of different
branches of science have been yearly drawn up and
circulated. Sums of money from the Association funds
have enabled committees and individuals to pursue
scientific researches which they could not otherwise
have done, and it has successfully recommended to
Government from time to time worthy scientific purposes and expeditions, which required grants of money
beyond its own means. It cannot be doubted also that
rewards of science were much more freely bestowed
after this popular agitation. Up to the year 1830, not
one title had been conferred upon men of science, but
between that time and 1850 twenty philosophers and

K

authors received knighthood ; thirty scientific and literary persons received new pensions, and seven members of the British Association, all of high scientific reputation, were appointed to lucrative and honourable posts. It was not only in the origin and objects of the British Association that Brewster strove to advance the long-neglected interests of science. The different footing and the higher position that men of science had occupied in France since the days of Colbert, who with true wisdom brought the light of science to advance and to illuminate the practical work of administration, was ever present to Brewster's mind in humiliating contrast ; for many years he scarcely ever wrote a review, book, or pamphlet without introducing the subject with persistent ingenuity, and in most forcible language. He was always a consistent though moderate Liberal, but he considered the interests of science as no party question ; and so fearless and plain-spoken indeed were his attacks, that he was looked upon by both political parties more coldly than might otherwise have been the case. Still he accomplished much that he hoped for. Such burning words as the following could not fail to have effect on the public and official mind :—

"But it is on higher than utilitarian grounds that we would plead the national endowment of science and literature. In ancient times, when knowledge had a limited range, and was but slightly connected with the wants of life, the sage stood even on a higher level than the hero and the lawgiver ; and history has preserved his name in her imperishable record when theirs has disappeared from its page. Archimedes lives in the memory of thousands who have forgotten the tyrant of Syracuse and the Roman consul who subdued it.

The halo which encircled Galileo under the tortures of
the Inquisition extinguishes in its blaze even the names
of his tormentors; and Newton's glory will throw a
lustre over the name of England when time has paled
the light reflected from her warriors. The renown of
military achievements appeals but to the country which
they benefit and adorn. It lives but in the obelisk of
granite, and illuminates but the vernacular page. Sub-
jugated nations turn from the proud monument that
degrades them, and the vanquished warrior spurns the
record of his humiliation or his shame. Even the traveller makes a deduction from military glory when he
surveys the red track of desolation and of war; and the
tears which the widow and the orphan shed, obliterate
the inscription which is written in blood.

" How different are our associations with the tablet
of marble or the monument of bronze which emblazon
the deeds of the philanthropist and the sage! Their
paler sunbeam irradiates a wider sphere, and excites a
warmer sympathy. No trophies of war are hung in
their temple, and no assailing foe desecrates its shrine.
In the anthem from that choir, the cry of human suffering never mingles, and in the procession of the intellectual hero ignorance and crime are alone yoked to his
car. The achievements of genius, could the wings of
light convey them, would be prized in the other worlds
of our system—in the other systems of the universe.
They are the bequest which man offers to his race, a
gift to universal humanity,—at first to civilisation, at
last to barbarism.

" Are these the sentiments of the statesmen of England, or have they ever struck a sympathetic chord in
the hearts of her people ? The hero, and the lawyer,

and the minion of corruption, and the truckler to
power, have hitherto reaped the rewards of official
labour, and usurped the honours which flow from the
British Crown. England alone taxes inventors as if
they were the enemies of the State ; and, till lately, she
has disowned her sages and her philosophers, and
denied them even ,the posthumous monument which
she used to grant to the poets whom she starved. It
is a remarkable event in the history of science, that in
1829, in one year, England should have lost Wollaston,
Young, and Davy, three of the most distinguished men
that ever adorned the contemporaneous annals of any
country. All of them had been Foreign Associates of
the Institute of France, all of them Secretaries to the
Royal Society, all of them were national benefactors,
all of them were carried off by a premature death, all
of them died without issue, and all of them have been
allowed to moulder in their tombs without any monumental tribute from a grateful country. It is not
merely to honour the dead or to gratify the vanity of
friends that we crave a becoming memorial from the
sympathies of an intellectual community. It is that
the living may lay it to heart, that the pure flame of
virtue may be kindled in the breasts of our youth, and
that our children may learn from the time-crushed
obelisk and the crumbling statue that the genius of
their fathers will survive even the massive granite and
the perennial brass.

"If we have appealed in vain to the sentiment of
national honour, to which statesmen are supposed to be
alive, we would now urge the higher claims of justice
and of feeling. If you are the minister of the Crown,
the dispenser of its honours, and the almoner of its

bounty, are you not bound by the trust which you hold
to place the genius of knowledge on the same level
with the genius of legislation and of war, to raise it to
the offices which it can fill, and reward it with the
honours which it has achieved? If the inventor swells
the national treasury, adds to the national resources,
strengthens the national defences, and saves the national
life, is he not entitled to the same position as those
who speak or who fight in the nation's cause? If
mercy is the brightest jewel in the royal diadem, justice
is the next; not the justice that condemns, but the
justice that recognises national benefits, and rewards
national benefactors. If the charge against England,
that 'she is a nation of shopkeepers,' is justified, as has
been alleged, by her disregard of intellectual pre-emi-
nence, we would counsel the ministerial head of the
firm to use just weights and keep accurate measures."[1]

When, after his death, it was stated "that the improved position of men of science in our times is chiefly
due to Sir David Brewster"—it was not more than the
truth. It may be mentioned here that one of the last
business acts of his life was to petition the Premier of
the Conservative Government of 1867 in behalf of the
widow and children of an early deceased man of science.
To those who knew all the circumstances of Brewster's
long conflict in behalf of his peers, it was deeply gratifying that the last act of Lord Derby's Ministry was
a prompt and favourable response, although it never
reached the ear that was cold in death before it was
received.

The success of the British Association was always
dear to Brewster; he attended most of its meetings, at

---

[1] Quoted from the *North British Review*.

one of which he made his last public appearance, and
in connexion with the Baconian principle of the " circuits or visits to the principal cities of the kingdom,"
I may mention that it is affirmed, and is probably the
case, that a casual remark of his on the subject suggested the Evangelical Alliance, which met for the first
time at Liverpool in 1845, and has continued its " circuit" ever since.  His name is on the original provisional committee of its promoters.

In 1831, the King (William IV.) sent the Hanoverian
Order of the Guelph to Dr. Brewster, Mr. Harris
Nicholas, and several other eminent men.  After their
acceptance in society as titled knights, it was discovered
that the Order conferred no title ; and an offer of ordinary knighthood speedily followed—a slender distinction, which Brewster was very indifferent about, and as
the fees amounted to £109, which found their way into
the pockets of the inferior servants of the Court, he
positively declined it.  It was still pressed upon him,
however, and he was informed that the question of fees
would be waived.  He therefore consented to go to London, and with his friend Mr. Nicholas went to the levée,
with the customary words " To be knighted" upon their
cards.  On presenting them, however, the lord in waiting
exclaimed that he knew nothing of it—a previous intimation to him having been forgotten.  An awkward moment
ensued, but my father said quietly, " Let us move on ;"
which was answered by the King's exclamation, " No !
no ! I know, I know !"  Having no sword, he borrowed
the Duke of Devonshire's, and with that performed the
usual ceremony.  The fees were never demanded.  An
anecdote was told at the time of some Waterloo officers
who, when the bills were sent in for their hardly earned

honours, took the accounts to the next levée, and left them on the King's table.

About this time his busy pen produced a *Treatise on Optics,* published in Lardner's *Encyclopædia;* he wrote his first short and popular " Life of Sir Isaac Newton," published in Murray's *Family Library;* and he also wrote a very popular work, which was published in the same series, " Letters on Natural Magic," suggested by and dedicated to Sir Walter Scott, and forming a useful corrective companion to his volume entitled, *Demonology and Witchcraft.* Scientific explanation of curious facts, which had hitherto been turned to the uses of superstition, was ever after a favourite subject, and a constant cause of correspondence. All sorts of optical illusions were communicated to him by letter from every quarter, which he patiently considered and answered. Out of these communications and his own observations he collected material for a second volume of Natural Magic, which unfortunately other work always impeded, and it remains unwritten. A review long afterwards, in the *North British,* of the work on the Occult Sciences by M. Eusèbe Salverte, forms a very interesting compendium of his views on this subject.

In this year an accident happened which nearly robbed Sir David of that wonderful and valuable eyesight which lasted to the end of his fourscore and six years. I well remember being waked from sleep in the middle of the night by a loud outcry, and the sight of my father, with outstretched arms, blinded and disfigured, rushing to plunge his head into the first basin of water he could reach. While pursuing his midnight experiments, a chemical substance had exploded right into his face. For many weeks he lay helpless, with bandaged eyes and disabled hands, a severe trial for his active temperament; but, from the following letter, he seems to have cast off the consequences, with that wonderful power of restoration which belonged peculiarly to his constitution. Some years before he had received a series of remarkably intelligent letters from Rome, with only the signature $\varDelta$, seeking information upon the most abstruse subjects of science. Much struck by these letters, written by a very young man, but full of mature intelligence, Brewster answered them fully and unreservedly, out of the stores of his own treasure-house of scientific knowledge. The anonymous signature continued for some time, and the incognito was not dropped till the writer's return to Scotland, when it transpired that he was Mr. James Forbes, a younger son of Sir William Forbes, Bart., of Pitsligo. He became a most intimate and frequent correspondent and visitor at Allerly, and thus wrote to Lady Brewster :—

"GREENHILL, 8th Jan. 1832.

" MY DEAR MADAM,—Your last kind letter was most welcome to me. From it and the one which Mr. Robison received the other day, I was delighted to find that Dr. Brewster is well enough to resume his experiments. I entreat you to prevent his overworking his eyes, and not to suffer those " hissing gases " to which you allude. He must find some other way of performing his experiments. I hope the yellow acid which I begged Dr. Reid to send answered the purpose. We have great reason to be thankful that the accident was not so very much worse, as it might easily have been. A statement of it has reached the newspapers, which is, however, very correct, and I am asked on all hands for

the latest news.  As Professor Necker of Geneva says,
 the Doctor's eyes are not his own property, but belong to
the world.'  Graham's Island has actually disappeared,
and so suddenly that I should not be surprised if it
made a re-ascension.  Dr. Davy wrote me that he
thought it was based on clay-slate, which renders this
more probable.  There are excellent accounts of Sir
Walter's health[1] and comfort at Malta.  Sir W. writes
that he means to make a poem on Graham's Island, to
the tune of ' Molly, put the kettle on !'

"  . . . I am daily looking with interest for Babbage's
letter which you mention.  Trusting to Dr. Brewster's
improvement, I have addressed the accompanying letter
to him, though I need hardly say it is equally open to
you.—Believe me, with great regard, my dear Madam,
yours most sincerely,          JAMES D. FORBES."

Although unable to resist his dearly beloved experiments, the state of my father's eyes probably required
caution in writing, for we do not find so much literary
work as usual at this time; though he seems to have
taken a most active part in plans for the improvement
of the neighbourhood—the building of a suspension
bridge across the Tweed—the arrangements for Sir
Walter Scott's Monument, and similar occupations and
interests.

His circumstances were extremely embarrassed at
this time.  Having no private means, no regular profession, no remuneration from his inventions, and his
greatest literary undertaking having proved a complete
failure in a pecuniary sense, and with three sons to
send out into the world, his spirits often sank at his

[1] Sir Walter Scott.

prospects. His unfailing friend and college companion,
Lord Brougham, offered him a living in the English
Church at this time, and he seems seriously to have
entertained thoughts of accepting it. He corresponded
with the Archbishop of Canterbury on the subject, and
the Bishop of Cloyne offered to ordain him, as technical difficulties which existed in England, did not apply
to Irish ordination. The proposal, however, was ultimately declined, although on what grounds I do not
know. In 1833 a door seemed to open out of his
difficulties. The Chair of Natural Philosophy in the
University of Edinburgh became vacant by the death
of Sir John Leslie. He offered himself as a candidate,
and success would have appeared certain, as all the other
candidates withdrew, leaving him without a competitor,
except Mr. James Forbes. The contest, however, went
on, and was not settled by scientific precedence or
European fame ; it became a question of family interest
and political party, which, after a severe struggle, ended
in Mr. Forbes's appointment. This was perhaps the
most severe disappointment of Brewster's life. Although not caused by any zeal for science on the part
of the Town-Council of those days, it is satisfactory to
remember that the future eminence of the young professor fully justified the appointment, and more than
fulfilled the promise of his youth, while in a few years
the broken friendship was cemented, and, both personally and by correspondence, became closer than ever.
Their careers curiously touched, as Professor Forbes,
with Brewster's warmest co-operation, succeeded him
in the Principalship of St. Andrews, and although many
years younger, only survived him a few months.

In the same year he moved with his family to Belleville, the scene of much early enjoyment.  By the death of his brother-in-law, his wife's sister was now the proprietor, and wished for her relatives to cheer her solitary home.  The guidance of a heavenly hand we find thankfully recognised in a brief sentence in a letter to his wife :—" You see, my dearest Juliet, how the Almighty provides for us when man cannot and will not, and how our confidence in Him can never be misplaced."  His three years' residence at Belleville was a distinct and separate episode of his life, and while presenting new experience in many ways, was extremely characteristic. The reform of abuses, which was a passion of his life, came into full play.  An extensive, but unremunerative, Highland property, for many years too indulgently superintended, and now under female sway, presented a fair field of reform, and he threw himself into it with all the ardour of his disposition. Farms let to unworthy tenants were speedily purified and filled by better and soberer men; careless officials were sent to the right-about ; and a new reign of order and business habits inaugurated, under which trees were planted, waste ground reclaimed, and a water-course planned and executed.  Various abuses in the neighbourhood were also examined into, but of course not without some of the unpopularity which reformers ever encounter, and of which he always had a full share.  Sir David, however, awakened a warm and abiding attachment amongst the majority of the Highland tenantry, who anticipated with delight the time, which never came, when he might be their landlord in very deed.  They were proud of his scientific fame, which indeed spread far and near.  I remember four working men coming a considerable distance from Strathspey, with the petition that they might see the stars through his telescope, while on another occasion a poor man brought his cow a weary long journey over the hills, that the great optician might examine her eyes, and prescribe for her deficiencies of sight; and all, as was ever his wont, were received courteously, and had their questions not only answered, but answered so clearly and patiently, that the subjects were made perfectly intelligible and interesting. He took great interest in the election for the county when his friend Mr. Charles Grant (Lord Glenelg) was returned as Liberal member; one old man whom he had canvassed, and whose knowledge of "Sassenach" was limited, was particularly cordial in his promise of his vote, but when he arrived at the hustings it was with the firm determination to vote for " Sir David," and nobody else, which he stuck to manfully. In order to qualify himself as a voter, my father had purchased a little cottage in the village of Lynchat, which was occupied one happy summer by his Peeblesshire friends, Mr. and Mrs. Montgomery and their family. With the exception of the Cluny Macphersons at Cluny Castle—then, as now, a centre of cordial hospitality—and one or two resident families, there was little society in winter, but every summer brought an influx of " shooters" and gay " Southrons." The brilliant coterie of Jane Duchess of Gordon, and the milder influences of Lady Huntly, had passed from Kinrara, but it was rented by Sir George Sitwell and his family. Edward Ellice, M.P., was at Invereshie, and at the Doune of Rothiemurchus, instead of its proprietors, the Grants, who were intimate friends of the Brewsters and Macphersons, the late Duchess of Bedford, with her gay circle of fashion, of statesmen, artists, and lions of all kinds, produced a constant social stir, in which Sir David was frequently called to bear his part, and he retained many lively recollections and anecdotes of the strange scenes and practical jokes of that "fast" circle. Upon one occasion, he and Lord Brougham, when Lord Chancellor, were visiting at the Doune. Lord Brougham, being indisposed, retired early to rest one evening. An hour or two afterwards the question was raised whether Lord Chancellors carried the Great Seal with them in social visiting. The Duchess declared her intention of ascertaining the fact, and ordered a cake of soft dough to be made. A procession of lords, ladies, and gentlemen was then formed, Sir David carrying a pair of silver candlesticks, and the Duchess bearing a silver salver, on which was placed the dough. The invalid Lord was roused from his first sleep by this strange procession, and a peremptory demand that he should get up and exhibit the Great Seal; he whispered ruefully to Sir David that the first half of this request he could not possibly comply with, but asked him to bring a certain strange-looking box; when this was done, he gravely sat up,—impressed the seal upon the cake of dough,—the procession retired in order, and the Lord Chancellor returned to his pillow.

He was much interested in all the old tales and legends of the country, and took much pains in excavating a strange hollow, of which many clannish stories were told, but which turned out to be a Pict's house. The parallel roads of Glenroy, long believed to be the hunting roads of the old kings of Scotland, with the various geological solutions of the ancient mystery, were objects of vivid interest. The weird stories of the glen and forest of Gaick, and the traditions of "Old Borlam," a

Highland laird, with certain Robin-Hood views as to
the rights of *meum* and *tuum*, who had formerly possessed Belleville, were repeated by him with lively
interest,—the cave from which Borlam and his men
used to watch for travellers on the old Highland
road was always pointed out to visitors,—and he
used to give as an example of the primitive state of
society in the north, which would scarcely be credited
in the south, that he had himself been in society, during his earlier Badenoch life, with Mrs. Mackintosh of
Borlam, the brigand's widow, a stately and witty old
lady. One day she had called at Belleville, and took up
*Lochandhu,* a novel just published by Sir Thomas Dick
Lauder; "Ay, ay," said she, "and what may this be
about?" to the consternation of the Belleville ladies,
her husband's capture and robbery of Sir Hector
Munro of Navar, and her own assistance in this, his
last exploit, by picking out the initials on the stolen
linen, being graphically detailed therein! On another
occasion Sir David had met her at a ball at Kinrara (in
1819), when Prince Leopold of Saxe-Cobourg was quite
delighted with her quaint racy conversation. When
her " carriage" was announced, one of the Prince's
aides-de-camp stepped forward and offered his arm.
She hesitated a moment, and then said, with an air
of resignation, " Well, well, I suppose you'll have to
see it." He returned in fits of laughter, for the old
lady's carriage was a common cart, with a wisp of straw
in the middle for a seat.[1]

[1] The account of his knighthood, and several of these Badenoch anecdotes,
are taken from jottings of my father's conversation when he first visited his
son and daughter at Belleville in 1862, which were taken down at the time
by Mrs. Macpherson.

A sudden flood of the Spey, as was its wont, came up during one of the summers of my father's residence at Belleville, over the flat meadows in front of the house, in which are the three small lochs called the Lochandhu, from their strangely dark aspect even under sunny skies. This particular flood scooped out a circular hole of great depth, which has remained ever since filled with water. The contents of the ground had been thrown out, and from their examination and other proofs he considered that there must have been a greater number of successive forests buried there than anywhere else in the known geological world.   Exactly thirty years later, when visiting his son at Belleville, the railway cuttings were going on, and he was keenly interested in verifying his former statement.

The glories of the Grampian scenery contributed more than anything to the enjoyment of his residence in Badenoch.   The beauties of the Doune, Kinrara, and Aviemore, Loch-an-Eilan, Loch Insh, Loch Laggan, Craigdhu, the Forest of Gaick, and the magnificent desolation of Glen Feshie, were all vividly enjoyed by him with that inner sense of poetry and art which he so pre-eminently possessed.   His old friend, John Thomson, the minister of Duddingston, but better known as a master in Scotch landscape, came to visit him, and was of course taken to see Glen Feshie, with its wild corries and moors, and the giants of the old pine forest.   After a deep silence, my father was startled by the exclamation, " Lord God Almighty !" and on looking round he saw the strong man bowed down in a flood of tears, so much had the wild grandeur of the scene and the sense of the One creative hand possessed

the soul of the artist. Glen Feshie afterwards formed
the subject of one of Thomson's best pictures.

In 1836, Dr. Brewster and his family left the Highlands, finding a residence there in many respects incon-
venient, and after passing some months in Edinburgh
they again took up their abode at Allerly.

## CHAPTER XI.

### NOTES OF LIFE FROM 1836 TO 1844.

Dwells there a shade on each lofty brow?
Falls there a tear o'er the severing vow?
Each eye like a falcon's is flashing bright,
Each brow is calm, and each step is light.

They have knelt at the Father's triune throne,
And they know they are His, "they are not their own,"
And onward they go—though each hope hath fled,
From an earthly sceptre,—a crown-wreathed head!

They go, and the lip of the scorner may curl,
His sword may flash forth and his flag may unfurl,
But blessed, thrice blessed, their path shall be,
They have sprung from their fetters! Their Church is free!

THE sixth meeting of the British Association took place at Bristol, commencing on the 22d of August 1836. Sir David Brewster went south to attend this meeting, which was a very interesting and successful one, under the presidency of the Marquis of Lansdowne. The week previous he spent at Lacock Abbey, the residence of H. Fox Talbot, Esq., the distinguished inventor of the Talbotype or Paper Photography. Professor Whewell, Charles Babbage, Esq., Sir William Snow Harris, Professor Wheatstone, Dr. Roget, and other men of eminence were assembled in a memorable group, and they all went together to the meeting at Bristol—my father visiting Mr. Daniell at Clifton. He wrote to his wife :—

"LACOCK ABBEY, CHIPPENHAM,
*Aug.* 15, 1836.

"MY DEAREST JULIET,—On my arrival here a few hours ago, I found a letter from Lord Fitzroy Somerset,

L

announcing Henry's promotion, which I have sent to him by this day's post. . . . This place is a paradise —a fine old abbey, with the square of cloisters entire, fitted up as a residence, and its walls covered with ivy, and ornamented with the finest evergreens. All are Whigs, and our only stranger to-day is Tom Moore, a most delightful person, full of life, humour, and anecdote.  He lives at a place called Sloperton Cottage, about four miles from this, and I hope in a day or two to have the pleasure of seeing him in his own house.

" *Aug.* 17.—In consequence of taking a ride to Bowood, the seat of the Marquis of Lansdowne, with Mr. Fielding, I was unable to send this by yesterday's post.  Bowood is the very perfection of art in landscape gardening, and is everywhere distinguished by the fine taste of its owner. . . . Our party was increased last night by Dr. Roget, Mr. Babbage, and Professor Wheatstone, so that we have all the elements of spending an agreeable week here.  Baron von Raumer is also to be at Mr. Daniell's, Clifton."

The art of Paper Photography, although it had been experimented upon by Mr. Fox Talbot since 1834, was not published to the world till January 1839.  It became a source of life-long interest to Sir David.  Mr. Fox Talbot sent him many of his earliest designs in photography—lace, leaves, printed pages, and picturesque bits of the old cloisters, and although much faded, these are still carefully preserved as interesting to compare with the present degree of perfection to which that wonderful invention has been brought.  My father's connexion with photography and photographers might

well furnish a chapter of his life in competent hands.
A large correspondence was kept up with Mr. Fox
Talbot, M. Claudet, Mr. Buckle, Paul Pretsch, Messrs.
Ross and Thomson, and other eminent photographers.
He made many experiments in the art, though not able
to give sufficient time to master its difficulties.   His
youngest son, when at home on leave, practised it under
his superintendence, and it was one of his father's
means of relaxation from heavier work, to take positives
from the negatives of his son and others.  A new photograph was to the last a joy to him, and he was peculiarly pleased with the receipt of a medal from the
Photographic Society of Paris in 1865.   I extract the
following touching account of the termination of his
correspondence with M. Claudet, the celebrated photographer in London, from the memoir of the latter by
his son : —

" Claudet's scientific relations with Sir David Brewster
had an affecting conclusion.   The two philosophers, for
some months during last year, 1867, were concurrently
engaged in investigating an interesting point in the
optics of photography.  The correspondence was broken,
never to be renewed, by the death of one.   The other,
sixteen years the senior, undertook to write a memoir
of his friend.   In a letter dated ' Allerly, Melrose,
January 1, 1868,' addressed to Mr. Frederick Claudet,
he says :—' . . . I shall be glad to do anything you
desire that can do honour to his memory, and I will
thank you to send me the fullest information in your
power respecting his early as well as his later life and
inventions.'  Six weeks later ' that old man eloquent '
passed away, and the full testimony he would have borne
to the scientific worth of Claudet—is not.  The chief sub-

ject of the letters of Brewster referred to, is the greater
perfection of photo-portraiture by means of small lenses
made of materials of different dispersive powers, with a
view to obtaining a depth of focus unattainable with
glass lenses. These letters are indeed surprising instances of vigour and freshness of intellect in a man
of eighty-six." [1]

After the meeting at Bristol, Brewster returned to
Allerly. Again the cares of pecuniary difficulties
pressed heavily upon him. He knew that at any time
he was liable to utter ruin should he lose the *Encyclopædia* lawsuit, and thus be exposed to heavy legal
expenses and accumulated arrears ; this period of
anxiety caused an irritability of nerves and of temper,
and a fear of poverty, which never again quite forsook
his finely-strung organization.

In 1836 the grant of £200 a year, in addition to
£100 which had been given previously, was made by
Government, and in 1838 the gift from the Crown
of the Principalship of the United College of St.
Salvator and St. Leonard, in the University of St.
Andrews, finally relieved him from all these embar-
rassments, which never occurred again to any serious
degree, although the old apprehensions were apt to
return from any fresh pressure of *Encyclopædia* claims,
and from a certain want of proper proportion in the
expenditure of his income. His appointment took
place in January ; he moved to St. Andrews in February, was inducted on the 6th of March, and on the
6th of April took possession of the old house, which he
had purchased, called St. Leonards, which was to be his

[1] "A. Claudet, F.R.S." A Memoir reprinted from the *Scientific
Review.*

home of joy and sorrow, of many changes and much
ardent work for twenty-three years. The old house had
formed part of the building of the ancient College of St.
Leonard, and had been the residence of George Buchanan, the old reformer and the stern tutor of James VI.
During the time of the Reformation, when any one
was supposed to be tainted by the new heresy, it was
significantly hinted by his friends that " he had drunk
of St. Leonard's Well." A pure reservoir so called is
still found near the College. It was a gloomy-looking
residence at first, with its arched gateway and its old
chapel, containing several tombstones, just in front of
the entrance-door, but it soon assumed a cheerful
and comfortable appearance, with its tiny lawn and
garden, and its creepers of ivy and jessamine, fuchsias
and roses. In the chapel was interred a predecessor
in the principalship, the same John Rutherfurd who
had received his education at an early date in the
Grammar School of Jedburgh, while the grave of Samuel
Rutherford, another Jedburgh worthy, is not far off in
the Cathedral cemetery. The other and the larger part
of the old College of St. Leonard was occupied by the
late Sir Hugh Playfair, Provost of St. Andrews, a man of
great eccentricity, unbounded energy, and real talent.
The close neighbourhood and some similarity of temperament occasionally produced clouds in the horizon,
but there was mutual warm regard besides a degree of
scientific sympathy, especially in photography, leading
to a constant intercourse, which was on the whole a
source of great interest to both. Sir Hugh Playfair died
in 1861.

Probably about this time, though I do not exactly
remember the year, another serious threatening of mis-

chief to Brewster's precious eyesight took place, causing him much anxiety and distress. Weakened probably by the accident at Allerly, his eyes were nevertheless more tried than those of ten ordinary men, not only by constant reading and writing, but by gazing through mysterious "bits of glass" at noonday, and by microscopic and other experiments by gas-light. An acute and agonizing pain suddenly darted into his eye-balls, deluging them with water, and necessitating complete darkness and quietness till the paroxysm had passed, which was sometimes not for two or three days. This complaint recurred frequently, and yielded to no mode of treatment, till at last he heard accidentally of a cure said to be discovered by Sir Benjamin Brodie, which consisted in using three or four times a day in the ordinary way, common snuff mixed with powdered quinine in equal proportions. This had a most rapid and wonderful effect, and he never again appeared to have any weakness or suffering in his eyes, although to the last he never spared them; in some of his optical writings, however, he alludes to having had symptoms both of hemiopsy or half-vision, and also of incipient cataract. Some years after, on mentioning the good he had derived from this prescription to his friend Sir Benjamin Brodie, its supposed originator, he found that the latter had never heard of it, and was much surprised by the effects, although he admitted the possibility of the cure, supposing the disease to be neuralgia.

In 1837 Brewster published a *Treatise on Magnetism*, originally written for the *Encyclopædia Britannica*, and in 1841 he found leisure to give to the world one of his most popular works, *The Martyrs of*

*Science,* being the biographies of Galileo, Tycho Brahe, and Kepler. The significance and quaintness of the title excited much pleasantry, and a circumstance that occurred in connexion with it long formed a favourite element in the pleasant household raillery which my father was so pre-eminently good-humoured in sustaining and enjoying. To the author's surprise and horror he found the following item in a "Christmas Box" which was handed to him :—"For binding four Martyrs, so many shillings !"

In the summer of 1842 Brewster took his wife to Leamington, to try the advice of the great "magician of the Leam," Dr. Jephson, for her long failing health. That remarkable man and the Scotch philosopher were mutually attracted, and they enjoyed frequent and genial intercourse. Leaving the recruiting invalid under his kind care and that of Scotch friends, Sir David took his daughter to the twelfth meeting of the British Association, held at Manchester under the presidency of Lord Francis Egerton. It was pleasant to see the honour and distinction which attended him. "There he is—that's Brewster !" were constantly-heard whispers, and it was a well-filled hemisphere in which he moved as a star of the first magnitude. One feature of the increasing success of the British Association has ever been the numbers of men of science from other countries who have come especially to attend these great gatherings. Upon this occasion there was a pleasant mingling of all nations, and a few amongst the number were Herschel and Bessel, the representatives of English and Prussian astronomy ; Sir William Rowan Hamilton, Dr. Lloyd, and Professor Maccullagh (whom my father termed "the three leaflets of the Irish shamrock") ; Pro-

fessor Jacobi, and M. Ehrenberg, his distinguished son-
in-law; Whewell, Murchison, Fox Talbot, Sedgwick,
Scoresby, General Sabine, and Dr. Dalton, fondly called
" the father of science in Manchester." This venerable
man appeared bowed down by age and infirmities,
which prevented him from presiding at this meeting,
and it was his last appearance at a British Association,
but wherever he was seen he excited much interest.
Brewster had an especial admiration for him, and a few
years after reviewed his memoirs and works, saying of
him that " among the great men who have illustrated
the passing century, there is no brighter name than that
of John Dalton." The peculiarity of vision which characterized this venerable philosopher, of which little
was known for a long time, was called Daltonism before
it received its unpronounceable name of Chromato-
pseudopsis, or, as it is now simply called, Colour-blindness. Dr. Dalton's inability to distinguish red from
other colours was supposed to be the cause of his occasional choice of a costume unusual for any, especially for
one like himself, belonging to the sober-habited Society
of Friends. When he and Brewster, along with some
other men of science, received the honour of D.C.L., during the meeting of the British Association at Oxford in
1832, Dalton was the only one of the group who wore his
scarlet robe all through the week, and two years later
he attracted general attention by appearing in the same
gay colouring at a Court levée. The subject of colour-
blindness was one of the many which Brewster took
up with vivid interest. Although not the first to bring
it before the public, his notices of the subject in
*Natural Magic* and in his *Treatise on Optics*, drew more
attention to it as an interesting and important optical

inquiry.   Several of his friends besides Dr. Dalton
had this imperfection : his old professor, the eminent
Dugald Stewart, Mr. Troughton, the astronomical in-
strument-maker, and others.   He examined many cases
of colour-blindness, gathered fresh facts and anecdotes, both by correspondence and conversation, and
wrote an interesting article in the *North British* on the
works of his friends the late Dr. George Wilson and
Professor Wartmann of Geneva.

From Manchester we went to Cambridge for another
brilliant week, living in the rooms of Professor Potter,
which, being vacation time, he kindly vacated for our
use in Queen's College, once an old Carthusian convent.
It was the occasion of the installation of the Duke of
Northumberland as Chancellor of the University, and
to nearly all the celebrated names which had assembled
at Manchester, were added Lord Rosse, Monckton
Milnes the poet (Lord Houghton), Buckland, Sir Mark
Brunel, Hallam the historian, the Duke of Wellington,
and many others ; the conferring of degrees by the new
Chancellor upon all the eminent men, who had not
already received them, was part of the interest of the
occasion.   Many memorable gatherings in Senate Hall,
Colleges, and Gardens took place, but best remembered
of all is the fine statue of Sir Isaac Newton, seen for
the first time by moonlight, with his biographer and
loving disciple, amidst the solemn beauty of King's
College Chapel.   From thence we went to the Deanery
of Ely for some days, to visit Dr. Peacock.   Our travelling companions were Dr. Buckland, famous for wit
as well as geology, and Professor Maccullagh, one of the
most brilliant mathematicians of his day, and of a truly
reverent and Christian mind, although, like Buckland's,

not very long after that happy visit, it completely gave way.

The spring of 1843 was too memorable a time for us in Scotland, and too decidedly an era in my father's life to be passed over in silence.   Lay patronage had always been considered a grievance by the evangelical section of the Church of Scotland, and had, for nearly a century, been rigidly administered by the Church Courts. It had already caused the secession of the Burgher and Relief Churches, and was now presenting its worst aspect in many parts of the country.    Being supported by the legal courts, several forced settlements of unwelcome and unfit pastors, especially those of Marnoch and Auchterarder, hurried on a crisis which an evangelical majority of the Church of Scotland had for some years been striving to avert.    Their efforts produced what has been called "the Ten Years' Conflict"—a conflict terminated in 1843.    On the 17th of November 1842 there was a solemn convocation held in Edinburgh, at which 465 ministers took their places.    A memorial was prepared and addressed to Government, in which it was calmly and clearly stated that the inevitable consequence of a continued refusal of relief must be a retirement from their position as connected with the Establishment, rather than the continuance of an unseemly contest.    On the 18th of May 1843, at noon, the Rev. Dr. Welsh, the Moderator of the preceding Assembly, preached the sermon before the Queen's Commissioner and the public in the ancient church of St. Giles, which, as is the wont of the Scotch Church, always precedes the meeting of the General Assembly, and chose for his text these words, " Let every man be fully persuaded in his own mind " (Rom. xiv. 5).    At half-past two o'clock the Assembly met in St. Andrew's Church, which was crowded from floor to roof. After earnest prayer, Dr. Welsh read a solemn protest of the Church of Scotland by her commissioners, against the oppression of the civil power, which had been signed the night before in St. Luke's Church by 203 representative ministers and elders, in which document the approaching event was styled " our enforced separation from an Establishment, which we prized and loved, through interference with conscience, the dishonour done to Christ's crown, and the rejection of His sole and supreme authority as King in His Church." Dr. Welsh then laid the protest on the table, bowed to the representative of Majesty, and left the church, followed at the time by 203 ministers (a number speedily increased to 474), and many elders, with their protesting adherents who had gained admission into the building.

.That upwards of 400 ministers should resign manses and stipends, or status and prospects, at the mere call of conscience, was a thing so little in accordance with the fashions of the nineteenth century, that the long and solemn warnings of it were treated as fiction. It seemed as if the eyes of statesmen, officials, and clerical advisers were holden, that they might not see the inevitable truth. Irreverent jokes, bets, and satirical prophecies were circulated through the country, and men had made up their minds that they were not to be shamed by the sight of an old-world triumph of principle. But when the words " They come ! they come !" thrilled through the hearts of the bystanders, announcing the solemn FACT, and when, arm in arm, the protesting men, with firm faces, but many with aching hearts, walked out into the streets of their Scotch metropolis, then the prophets *had* honour in their own country.    The pulse of the nation was stirred.    Mind triumphed over matter, soul over flesh, conscience over mammon, and the gazing thousands of the city were moved into tearful admiration.    When the fact reached the ears of Lord Jeffrey in his quiet study, in surprise and incredulity he asked the universal question, "*How many?*" and when the answer came, he burst into tears, exclaiming, "Thank God,—in no other country could such a deed be done!"

On Tuesday the 23d, in Tanfield Hall, Canonmills, the protesting ministers signed the Deed of Demission. It was a noble sight—one of the solemn joys of a lifetime to witness,—as the excitement over, each brave man took his pen and irrevocably signed away home and income.    There were additional signatures also, which were peculiarly noticeable and valuable, for they were those of men who had wavered on the day of Disruption, perhaps because of the tears of their wives, and the wails of their children.    Yet conscience, enlightened by the word of God, had done its sure work, and with judgment cool and collected, they came forth to the place of signature with their feet planted on the promise, "The Lord will provide."    And He DID provide.

David Brewster had taken part in every step of the long conflict—he signed the Act of Protest, where his well-known writing is still shown; with his elder brother James he walked in the solemn procession, and he attended every sitting of that first Assembly of the Free Church of Scotland, the opening psalm of which was emblematic of her future, for as the words "O send thy light forth and thy truth,"—led by the magnificent voice of Mr. Hately, and then taken up by 4000 singers,—echoed through the pointed roofs of the spaci-

ous Hall of Canonmills, a bright beam of heaven's light
shone out and dispelled the thick darkness of a previous thunder-storm.   It was not excitement that caused
his secession from the Church of his fathers, and of his
own deeply-rooted attachment.  Excitement, political par-
tisanship, deep, earnest indignation at refused sites and
petty persecutions, and some of the "madness" which
"oppression" causeth even to "wise men," there was
much of in the mind of the day, fostered by the brilliantly
witty and vehement articles of the *Witness*, the principal
organ of the Free Church, edited by Hugh Miller, the
eminent geologist, and the *Fife Sentinel*, a less known
but equally vehement paper, edited by David Maitland
Makgill Crichton of Rankeillour Makgill, in both of
which publications Brewster took part, sending notes
and hints for articles when not able to write himself,
taken from the books he read, or the conversations he
shared.   His friendship with these two gentlemen,
indeed, probably kept up much of his church feeling,
and every Christmas regularly for many years he and
Mr. Miller spent some days together at Rankeillour,
where geology and ecclesiastical topics reigned supreme.
But the progress of events would alone have prevented
the decay of party feeling.   The Free Church was not
without her martyrs; sites refused for churches and
manses in which men might worship and dwell with
a free conscience, did something of the work of sterner
instruments in days of old.   Some of the Disruption
ministers were infirm, others aged, many peculiarly
liable to the various ills to which flesh is heir, and for
such preaching in gravel-pits—in ships—on the high-
ways—on the shore below high-water mark—travelling
many a Highland mile in storm and tempest, or sleep-

ing in attics under the drip of the rain, could not be
conducive to health, strength, or life. Many con-
tracted disease, and some lay down and died. It
was, however, the ministers, their wives, and their
children, who suffered in this fashion. The elders of
the Free Church, though undergoing much social in-
convenience, were, as a body, free from loss or suffering.
One man was, however, called at the period, *par excellence*, "THE suffering elder of the Free Church." That
man was David Brewster. It appeared that he was the
only official in a similar position who had "come out,"
though others were supposed to have had equal desire
though not equal courage. In 1844, therefore, proceedings were commenced against Sir David by the
Established Presbytery of St. Andrews, aided by the
University, to eject him from his chair as Principal, because of his adherence to the Free Church. The Test Act
was made much of—an Act instituted originally to keep
out Episcopalians, several of whom were calmly occupying, at the very time, Scotish Professorial chairs without
remark or question. Amongst these reverend and
academic gentlemen, only one was found bold enough
to take the part of the heretical Principal, the late
Rev. Professor Ferrie. Public opinion, however, was the
best defence in such a case, and after months of small
attacks and annoyances, and irritating summonses,
which it must be confessed were not borne with equa-
nimity, and which much deepened party prejudice in
my father's mind, the proceedings were at last dropped,
technically, I believe, because he had not signed the
formal Deed of Demission, which no elder had done.
The following short account he wrote to his wife :—

" My case was *quashed* in the Residuary Assembly.
They durst not look it in the face, and therefore gave
the decision the appearance of having been only delayed.
Dr. Ferrie objected to the word '*meanwhile*,' which
indicated that it was not at an end, but Dr. Mearns, the
moderate leader, begged of him to say nothing about
that, as this was '*their way*' of getting rid of it altogether!"

But when the excitement and persecution was all
over and gone,—when again, as in the old days, he had
warm friends among Established Church ministers, and
occasionally worshipped within her pale,—when he had
seen the worst and the best of Free Church government,
he still held that though not perfect, it was the purest
and nearest the Scriptural Church, and maintained,
with the calmest, strongest judgment, the principles of
the Protesting Church of Scotland, *i.e.*, the spiritual
independence of her Courts, and the right of her people
to choose their pastors. Wherever he went he fought
her battles ; and when in England,—amazed and half
amused by the profound ignorance existing on the
subject, even amongst thinking minds,—he was accustomed to recommend a book which he thought gave
the most clear and incontrovertible statement of the
truth. It was entitled *The Scotish Church Question*,
written by the Rev. Adolphus Sydow, chaplain to
the King of Prussia, by whose private desire, it was
said, he came over to Scotland, studied both sides
of the question, and published his impartial opinion.
It was a grief to my father that only one of his immediate family belonged to the same communion as him-

self; and of one near connexion, whom he highly valued, he said, twenty years after the Disruption, " It CAN only be because he has not studied the subject; he must read Sydow." For many years his silver head was seen regularly at every meeting of the Free Church Assembly, and his correspondence shows that he left his science and his writing to make the most careful arrangements for supplying pulpits and getting candidates to be heard, attending also punctually Commission and committee meetings, of which the following undated letter gives an idea :—

" As Tuesday is a very busy day in the affairs of the Free Church, I have resolved to go to Edinburgh to-morrow in the train which arrives there at 3.40. There is a meeting of the Education Committee at ten; a conference with the United Presbyterian Committee at twelve; another meeting of the General Education Committee at 38 York Place, at one o'clock; a meeting of the College Committee at three, in the New College, George Street; and a meeting of Lady Effingham's Bequest Committee at seven o'clock in the evening. Of all these committees I am a member, and the subjects are of such importance that I feel it a duty to be present."

One kindred subject, although out of date, may be mentioned here. Like the large majority of the Free Church, my father was no voluntary. They did not leave the State till the State left them, and it was with extreme reluctance that they quitted an Established Church. But the tie once broken, in the case of many, it was so thoroughly severed that they began to see the blessings of a Church which did not " walk abroad in silver slippers," according to the saying of the old divine, and their desires for union have not therefore

gone so much in the direction of mending the tie
broken by the Disruption, as of uniting with the large
body of Presbyterians whose rules and worship are precisely the same as the Free Church, except that the
voluntaryism of the one was voluntary, while the voluntaryism of the Free Church was at the first compelled.
This view my father held with earnestness; his heart
was in what is called "the Union Question"—he
mourned over every delay, and three days before his
death he spoke of it as " the cause of God."

My father's early friendship with Dr. Chalmers, of
whom it was said at this time, "Where Thomas Chalmers is, there is the Church of Scotland," was not as
may be believed hindered, but rather furthered, by these
events. Much correspondence took place between
them upon Church affairs, and while there was yet
neither Free Church nor pastor at St. Andrews, Dr.
Chalmers became a guest at St. Leonard's College, and
preached in the open air to 4000 people in the green
amphitheatre between the sea with its far-stretching
rocks, and the monument to the martyrs who suffered
by fire at St. Andrews, which was then in the process
of being erected. A grand scene, and a noble sermon
on "Fury is not in Me." When Chalmers left St.
Leonard's, it seemed to those who had had the privilege
of receiving him, as if it had been an angel's visit,
so profound was the impression made by his child-
like humility, gentleness, and wisdom of speech.

M

## CHAPTER XII.

### NOTES OF LIFE FROM 1844 TO 1850.

THROUGH days of sorrow and of mirth,
Through days of death and days of birth,
Through every swift vicissitude
Of changeful time, unchanged it stood,
As if, like God, it all things saw,
It calmly repeats those words of awe,—
    " For ever—never !
    Never—for ever ! "
                            LONGFELLOW.

ABOUT this time it occurred to several gentlemen in
Edinburgh that "there was both room and need for a
Review of the highest class, the organ of no party,
political or ecclesiastical, and which, instead of ignoring
or affecting to disown Christianity, was imbued with
its spirit."[1] The *North British Review* was therefore
started in 1844, under the editorship of the Rev. Dr.
Welsh. The double coincidence of this event taking
place the year after the Disruption, and the Free Church
principles of most of its editors, led to the erroneous
impression in some quarters that it was the organ of the
Free Church. It has never been a sectarian work however; contributors of all denominations were welcome, if
their principles were good and their literary talent undeniable. The success of the undertaking was remarkable;
—from the very first this quarterly took the high place
in literature, which it has ever since sustained. It has
always been under most careful editorial superintend-

---

[1] Quoted from *Memoirs of Dr. Chalmers*, by Dr. Hanna.

ence of a high order—Dr. Welsh, Lord Barcaple, Dr.
Hanna, Professor Fraser, Dr. Duns, Professor Blaikie,
and Mr. Douglas, having successively managed it—
and the interest and variety of its articles have been
in proportion to the singular variety and eminence of
its contributors.    Up to this time Brewster had been
a contributor to Professor Napier's *Edinburgh Review*,
for which he wrote twenty-eight articles, but the establishment of this more congenial periodical was quite
an era in his literary life.    He threw himself into its
interests with the most cordial energy ; for upwards
of twenty years he contributed an article to almost
every quarterly number, and he delighted in beating
up for recruits for this service among his eminently
intellectual friends.    Professor Fraser writes,—" I have
many letters received from him during the time of our
literary connexion, when I was editor of the *North
British Review* in 1850 and the seven following years.
In that relation I always found him in the highest
degree kind, cordial, and considerate.    The freshness of
his nature was shown in his extraordinary readiness
to sympathize with the life and movement of the age.
He was among the most remarkable in a band of contributors which then included the ablest men of the
time in Great Britain, not only for the brilliancy and
vivacity of his writings, but for the punctual regularity with which they were delivered.    He contributed
an article to each number during the time I was editor,
and in each instance, after we had agreed together
about the subject, the manuscript made its appearance
on the appointed day with punctual regularity, and
its successive instalments were placed by him in the
editor's hands with mechanical precision.    Some of

the articles were the subject of interesting correspondence between us ; and I recollect in particular the ardour with which he addressed himself to the thoughtful and very suggestive essay on the *Plurality of Worlds*, which I had asked him to review, in an article since expanded into his *More Worlds than One.*" Professor Blaikie, who edited the *Review* from 1860 to 1863, writes,—" Sir David Brewster was ever remarkable for the carefulness of his work, the punctuality with which it was delivered—never behind time, never needing to write to the editor for more time or more space : a model contributor, indeed, in every way, and so full of well-put and attractive information. He was of great use in giving introductions to eminent men, his name being a guarantee that the channel in which they were asked to write would not be unworthy of them." The secret of the careful execution of this literary work was, that he spared no pains which could possibly perfect an article. Not contented with the book itself, which he had to review, and his own previous knowledge of the subject, he collected fresh information before beginning to write, from every source ; he was always specially anxious to obtain particulars of the life and career of the author, so that most of his articles possess a biographical value apart from the intrinsic interest of the subject. The variety is indeed most curious, as is best seen from the four thick volumes which I have before me, in which are collected all these contributions. Dr. J. H. Gladstone gave the following graphic description of some of these in his obituary notice before the Royal Society of London :— " The first number of the *North British* commences with an article by him, on Flourens's *Eloge Historique de*

*Cuvier;* and further on in the same part he discusses the *Lettres Provinciales* and other writings of Blaise Pascal. In the second number he describes the Earl of Rosse's great reflecting telescope; and shortly we find him engaged with such serious works as Humboldt's *Cosmos* or Murchison's *Siluria;* the rival claimants for the honour of having discovered Neptune divide his attention with Macaulay's *History of England*, or the *Vestiges of the Natural History of Creation.* With Layard he takes his readers to Nineveh, with Lyell he visits North America, and with Richardson he searches the Polar Seas. The Exhibition of 1851, the Peace Congress, and the British Association, come in turn under his descriptive notice; or, turning from large assemblies to individual philosophers, he sketches Arago, Young, or Dalton. In one number we have 'The Weather and its Prognostics,' and 'The Microscope and its Revelations;' elsewhere he describes the Atlantic Telegraph, whilst in a single article he groups together 'The Life-boat, the Lightning-conductor, and the Light-house.' He reviews in turn Mary Somerville's *Physical Geography*, and Keith Johnston's *Physical Atlas;* the History of Photography engages him at one time, and at another Weld's History of our Society. Under the guidance of Sir Henry Holland he investigates the curious mental phenomena of mesmerism and electro-biology, and under that of George Wilson he inquires into colour-blindness. He criticises Goethe's scientific works, expounds De la Rive's *Treatise on Electricity*, and Arago's on Comets; or, turning from these severer studies, he allows Humboldt to exhibit the 'Aspects of Nature' in different lands to the multifarious readers of the *Review.*"

His review of the *Vestiges of the Natural History of Creation* is one of the most remarkable, and I have heard that it was bound up with the book itself as an antidote, and thus sold in America.

It was one of my father's greatest literary pleasures to peruse carefully each number of the *Review* as it came out. For many years he had sent to him regularly, by the editors, a list of the authors of the different essays, delighting to copy and enclose it to friends at a distance, who he knew were readers of the work.

An incident connected with the *North British Review* caused my father so much interest, pride, and pleasure, that I cannot forbear mentioning it. The St. Andrews students had been left very much to their own devices for a good many years, and a state of things had set in which did not accord with the views of a new Principal of reforming tendencies. I believe that the breaches of discipline were not of serious importance, but quite sufficient to bring the University and its authority into some disrepute. Disorderly bands of red-gowned students patrolling the streets, too lengthened "gaudeamuses," unnecessary appeals to door-bells and knockers in the midst of the night, apparitions of tall, bearded guisards[1] into quiet families, and such like, were the principal offences, and with characteristic activity the perpetrators were brought before the Senatus, rebuked, and punished, so that a very different state of things soon came about, and St. Andrews students became as orderly as most of their class. A good many years afterwards my father read an article in the *North British*,

---

[1] A Scotish word borrowed, like many others, from the French, for children who go about masked, at Christmas and New Year, and sometimes act little childish scenes.

so fresh, so full of vigour and interest, that he at once wrote to Professor Fraser to know who the author was. His delight was extreme—indeed, I scarcely remember his ever showing more complete satisfaction—when he found that it was written by one of his old students of that somewhat stormy period, of whom he had never heard since. He at once wrote him a letter of congratulation on that happy beginning of a now long successful authorship, for the writer of the article was the Rev. John Tulloch, then minister of Kettins, but afterwards Principal of the Divinity College at St. Andrews, and author of *The Leaders of the Reformation, English Puritanism*, etc. The friendship between the young man and the old was ever after most cordial, especially during the years when they were contemporary Principals in the same University.

Sir David's interest in all his students was very great, and he was popular and accessible among them, having them at breakfast and tea as often as his busy mornings and experimenting evenings would allow of the interruption. This accessibility to young men, even though not his own students, was very marked. A friend writes,—" My personal recollections of Sir David are all of the most pleasant kind, and I gratefully remember his kindness to me as a student; he was ever ready to see me, and not only converse about my studies, but to tell me of his, showing me the nature of his discoveries, and performing some of his experiments. A more gentle and accommodating spirit to the young never glowed in a human bosom."

In the spring of 1845 my father paid one of his frequent visits to London, and we still find the habit carried on of frequent communication to his wife of all that he thought could interest her, although the extreme bustle of his London life renders his letters, written almost daily, difficult to make extracts from. The following are interesting :—

"MY DEAREST JULIET,—We have gained our cause on the question of Tests,[1] by the Government allowing Mr. Rutherford to bring in his bill, as you will see in the *Times*. Sir Robert Peel, attended by Sir James Graham, the Lord Advocate, and Sir George Clerk, received the deputation in his own house yesterday at twelve o'clock ; and though Sir J. Graham had the night before told Mr. Rutherford that the Government was to *oppose* the bringing in of his bill, yet the arguments and facts of the deputation, many of which I stated, from sitting next to Sir Robert, together with the eloquent and powerful speech of Mr. Rutherford in the House last night, prevailed, and the repeal of the Tests is now certain. From the House I went to the great meeting of the Protestant Delegates in Exeter Hall, where I found Guthrie on his legs electrifying a small audience of about 6000 persons ! He was followed by Baptist Noel, a most elegant and interesting-looking man, whose eloquence, chaste yet powerful, kept up the impression produced by Mr. Guthrie. We attended this morning a great public breakfast in the London Tavern, on the subject of the Maynooth Grant, which I think we shall yet defeat."

"LONDON, *May* 1, 1845.

" I have just time to give you my *Wellingtoniana*, which are rather interesting. . . . After a nice dinner-

[1] The Scotish University Tests were not done away till 1853.

party at Lord Rosse's yesterday, during and after which I had to fight the anti-Maynooth battle, as well as that of the Free Church, I hurried to the Archbishop of York's without joining the ladies. I found the great Duke seated beside Miss Georgina Harcourt on a double couch. As her object was to let me have some conversation with him, she soon summoned me to the empty side of the couch, and during nearly an hour I had the most unreserved conversation with the Duke and her. He speaks with a certain degree of difficulty, as if there had been a paralytic affection ; but not very perceptibly so. He amused us with his account of an American who wrote to him that he had come all the way from the United States to see him ; he did not, however, send him his address, for he said that if he showed himself in this way to one person he must do it to everybody, and this was impossible, unless he could be in more places than one at the same time. As Lord-Lieutenant of Hampshire, the British Association had asked him to be President at Southampton in 1846. Having seen his reply, I mentioned how much the Council regretted it, which led him to give us an account of the way in which all his time was employed throughout the year. From February till August he was obliged to attend the House of Lords, and never accepted an invitation on the days when the House sat. He often went home at twelve and one o'clock, and never got any dinner at all. His servant always asked him if he would have it, but at such late hours he preferred going to bed. From August till November he is obliged to live at the Cinque Ports, and, besides his military duties, he had occasionally to attend upon Her Majesty. For that reason he could not undertake the duties of President. We urged that attendance for one day would be sufficient, and that Lord Francis Egerton had done this at Manchester. This he could not do, as he would not undertake any duty without doing it completely.

" We then talked of Wheatstone and the electro-magnetic telegraph, and the conduct of the French in trying to introduce Wheatstone's inventions as a French system of telegraphs. He told me how the Duke of Buccleuch had, at the birth of Prince Alfred, got to Windsor before all the other Ministers, from the acci-dent of his servant having seen the telegraph at work. We talked much of Lord Rosse's telescope, the size of which he knew well; and upon telling him that the transit of Mercury was to take place to-morrow, and that a party was going to Sir James South's to see it, he said he would go, as he knew Sir James. We then talked of the new method of extinguishing fires by the sudden production of a great quantity of carbonic acid from charcoal, and of the Exhibition, which he had not seen, not having been able to go to the Royal Academy dinner. I mentioned the finest pictures, viz., Mr. F. Grant's picture of a Miss Singleton, and Sir W. Allan's *Nelson boarding the San Nicolas*, which he was to go and see some morning very early, to avoid the crowd."

" LONDON, *May* 9, 1845.

" We had a very singular scene at Sir James South's yesterday, where eight or ten telescopes were erected on his beautiful lawn at Camden Hill, to view the transit of Mercury over the sun's disc. The day was not good, but at four o'clock the clouds so far cleared away as to enable almost every person in the large party of fifty or sixty people to see the planet, like a round black patch, pass over the sun's face. The Duke of Wellington did not come, as he intended, being prevented by the dampness of the day, as he told Miss Harcourt, who was there."

Although undated, the following letter was written about this period :—

" MY DEAREST MARIA,—Upon coming here from the House of Commons (where we have just lost our University Test Bill by a majority of fifteen in a very full house), I have found your letter and its enclosures. We would not have lost it had it not been from the supineness of the Free Church and the citizens of Edinburgh, who sent up no deputation, and even no individual to enlighten and collect friendly and liberal members.

" Believing that the debate was to come on in the evening of to-day (Wednesday), I would have been able to do nothing in the matter, and not even to be present at the debate, had I not accidentally met with Sir Edward Colebrooke, who told me that the debate was to begin at twelve. I therefore hurried to the late Lord Advocate, Mr. Moncreiff, to put him up to several facts, I got one petition unkennelled from its place in the House, to which Mr. Ireland had improperly addressed it to Mr. Ellice, so that it was read before the debate. I conversed with a number of friendly M.P.'s in the Members' Gallery, Mr. Dennison, Mr. Philip Pusey, etc., etc., and prevailed upon them to stay to the vote.

" Don't be alarmed when I tell you that I was taken *prisoner in the House* by the Serjeant-at-Arms, Lord W. Russell, and *released* by order of the Speaker !    A

division was announced, and strangers ordered to withdraw.  I obeyed, went out of the House, but stood in
a corner of the stair, in place of going to the lobby, not
knowing the right thing to do.  When the division was
over I was taken prisoner, being found among the
members.  The Serjeant-at-Arms was puzzled, but having
got authority to release me, he called out through a
little window in the door that *his prisoner* was released,
and I emerged, to the amusement and amazement of a
number of members whom I knew, waiting outside for
admission.—I am, my dearest Maria, your affectionate
father,                                 D. BREWSTER."

"ATHENÆUM CLUB,
*Wednesday, 5 o'clock.*"

For two years his eldest son had been at home on
furlough from his duties in the Bengal Civil Service, and
on the 18th of September 1845 his marriage took place,
an event which brought into the family a peculiarly
gentle and lovely daughter, for whom my father ever
felt the tenderest affection.[1]  During the remainder of
their time in this country, before returning to India, the
newly married couple resided at Barham Lodge, within
a drive of St. Leonard's, which, with a large and pleasant
circle of new connexions, added much to his social en-
joyments.   St. Andrews itself contained much excellent
and intelligent society, in which, as well as in that of
a well populated country neighbourhood, he found that
pleasant social relaxation which he always needed from
study.   During part of his residence in St. Andrews,
he gave a series of popular scientific lectures, which he
made most attractive, as he was peculiarly fitted to do,

[1] Catherine Maitland, fourth daughter of James Maitland Heriot, Esq.
of Ramornie, Fife.

to crowded audiences of strangers and casual visitors, as
well as friends, acquaintances, and students.  About this
time Dr. Merle d'Aubigné, the eminent historian of the
Reformation, paid a visit to St. Leonard's College, which
was long remembered with the greatest interest.   Many
foreigners brought letters of introduction, one of the
most interesting of whom was M. Kossuth, the Hungarian
patriot ; there were also Count Krasinski, Prince Adam
de Sapieha, and several younger branches of the Genevese
families whose acquaintance Brewster had made in 1814.

In the spring of 1846 he went with his wife and
daughter to Rothesay for the mild climate and shelter
from east winds—enjoying keenly, as was his custom, the
new route, the lovely sea and island views, and pleasant
new acquaintances, and returning homewards in time for
the General Assembly of May, which was marked by
the laying of the foundation-stone of the Free Church
College, a ceremony which he attended with much
interest.   The breakfast party connected with this event,
and a long friendly visit just afterwards, were the last
occasions on which Brewster and Chalmers met on
earth.   One year after, in 1847, the former took part
in that mournful but wondrous procession which bore
Dr. Chalmers to his rest in the new Grange Cemetery,
of which it was written at the time—

> " A nation's blessings cheered thy living way,
> A nation's tears attend thy path to-day
> Where thronging multitudes beside thee tread,
> And severed creeds are meeting round thy bed.
> No mighty breather-out of song wert thou,
> No minstrel chaplet crowned thy massive brow,
> Nor conquering warrior on his bay-wreathed bier,
> Nor ruler of a realm nor ermined peer.
> No ! to thine honoured grave thou goest down,
> But locks of silver are thine only crown.

A lowly worshipper—a faithful pastor ;
A child-like servant of thy Heavenly Master ;
The pillar of thine own beloved fane ;
The wrencher of its chill and crushing chain ;
The master spirit of thy native land,
The famed—the loved in many a distant strand.
And now unconscious of each honour paid,
Thy noble brow must in the dust be laid ;
Thou goest, heedless of each heaving breast,
'Mid weeping thousands to thy place of rest.
Not where the mighty of the earth are sleeping,
Not where th' escutcheoned vault proud dust is keeping,
Near the green shade of trees we lay thy head,
Hallowing a new-made City of the Dead."

In 1847 Brewster was made a Chevalier of the Order
of Merit by the King of Prussia, whose acquaintance
he had before made at Taymouth Castle.    In the same
year he took his daughter and two friends, the Misses
Lyon, to the seventeenth meeting of the British
Association, which met for the second time at Oxford,
under the presidency of Sir Robert Harry Inglis, Bart.,
then the representative in Parliament of the University
—a peculiarly interesting locality for such a meeting,
which was moreover a very brilliant and valuable one
in all its component parts.    The twin discoverers of
Neptune, Mr. Adams and M. Leverrier, were present ;
as usual Brewster had entered keenly into the scientific
controversy which agitated French and English minds
as to the priority of this important discovery, so important as to extend the Solar system one thousand
millions of miles beyond its former known limits.    The
injustice that had been done to the young English
astronomer was only beginning to be repaired at the
Oxford meeting, when it was pleasant to see the two
distinguished and independent discoverers meeting on
equal and amicable terms.    The United College of St.
Andrews offered Mr. Adams their vacant chair of Natural

Philosophy, a compliment which had never before been paid to any one but Dr. Chalmers, in the case of the Moral Philosophy Chair in the same College.

In Brewster's *Life of Newton*, the following passage occurs :—" The honour of having made this discovery belongs equally to Adams and Leverrier. It is the greatest intellectual achievement in the annals of astronomy, and the noblest triumph of the Newtonian Philosophy. To detect a planet by the eye, or to track it to its place by the mind, are acts as incommensurable as those of muscular and intellectual power. Recumbent on his easy chair, the practical astronomer has but to look through the cleft in his revolving cupola, in order to trace the pilgrim star in its course ; or by the application of magnifying power, to expand its tiny disc, and thus transfer it from among its sidereal companions to the planetary domains. The physical astronomer, on the contrary, has no such auxiliaries : he calculates at noon, when the stars disappear under a meridian sun : he computes at midnight, when clouds and darkness shroud the heavens ; and from within that cerebral dome, which has no opening heavenward, and no instrument but the Eye of Reason, he sees in the disturbing agencies of an unseen planet, upon a planet by him equally unseen, the existence of the disturbing agent, and from the nature and amount of its action, he computes its magnitude and indicates its place. If man has ever been permitted to see otherwise than by the eye, it is when the clairvoyance of reason, piercing through screens of epidermis and walls of bone, grasps, amid the abstractions of number and of quantity, those sublime realities which have eluded the keenest touch, and evaded the sharpest eye."

From Oxford we went to Hartwell House, the fine old ancestral residence of Dr. Lee, which had been the abode of Charles X. and his family during their stay in England, after the memorable "three days" of July 1830. In this beautiful and scientific mansion were assembled a large party of the British Association, lions both home and foreign, and one of the many interests of this visit, peculiarly valued by my father, was the nightly observation of the heavens made in the noble transit-room of Hartwell.

In 1848-49 I find the following interesting notes of contemporary persons and events in home letters during his annual visits to London. The first letter refers to a period when M. Guizot, the eminent French statesman, after the Revolution of 1848, had made many inquiries about St. Andrews as a place of residence for himself and his family. A lengthened sojourn in this country was not however found necessary for the distinguished exile, and this plan was abandoned.

"LONDON, 1848.

"SIR HARRY VERNEY had called upon me and requested me to go at *three* o'clock to a public breakfast to the friends of *St. John's Schools*, leaving a card of admission for two. This most interesting establishment is sup-- ported by the zeal, and, to a considerable extent, by the wealth, of Mr. Arthur Kinnaird, Lord K.'s brother, a man of true piety. After visiting the Exhibition, Dr. A. and I went to the breakfast, or rather luncheon. Lord Ashley was in the chair, and Mr. Baptist Noel, beside whom I sat (and with whom I had some delight- ful conversation), and the Rev. James Hamilton, were among the speakers. The subject for conversation was that of providing amusements for the lower classes, and,

as ill luck would have it, Mr. Arthur Kinnaird asked
me to state my opinions, which compelled me to say
something, and in the course of conversation I was
obliged several times to say a few words, which I did
better than I thought possible.    I was then introduced
by Lord Ashley to Mme. Chabot, Mlle. Guizot, and
her brother, a boy of about thirteen or fifteen years
of age, who attended Adolphe Monod's church in
Paris.    Mme. Chabot is a pious woman, who has
impressed her own character upon the young Guizots,
and I should not wonder if, occasionally at least, they
should come to our church.    Mme. Chabot told me that
they (M. Guizot and his family) go to St. Andrews
early in August, and requested me to call upon M.
Guizot at 21 Pelham Crescent, Brompton, which I
mean to do to-day, and to give him all the information
he may require about our city.

" Lord Ashley mentioned in the course of his speech,
that he was that evening to open a reading-room in
Westminster, for the civilisation of the ragged and
thieving adults of that frightful locality.    I was anxious
to be present, and was appointed to meet him at the
House of Commons at half-past six.    Dr. A. and I went
there, and were joined by the Marquis of Blandford,
Lord Castlereagh, and others.    We walked to the place
of meeting, and first visited the Ragged School, taught
by a Mr. Aitchison, from Glasgow, quite a superior man.
The city missionary of the district is a Mr. Walker,
the son of a grocer (as he told me), at Earlston, near
Melrose, a most devoted man, and a man of great
physical energy.    The meeting for opening the reading-
room, and a room for teaching adults, was crowded to
overflowing, all the rogues of the place being either

N

inside or at the doors and windows. Lord Castlereagh and Lord Kinnaird moved two of the resolutions, and Lord Ashley spoke repeatedly. The behaviour of the people was admirable. The meeting went off in the most gratifying manner. It was delightful to see two young nobles who are to be, the one the Duke of Marlborough and the other the Marquis of London-derry, giving their time and their money for the ameli-oration of the condition of the poor. I felt self-reproached in considering how little we do in the same field of duty in our locality. After the meeting was over, and we had secured our silk handkerchiefs in our inner pockets, we went to the House of Commons, where Lord Ashley got us seats in the Speaker's Gallery to hear the discus-sion of the Sugar question, which was very uninterest-ing. Mr. Baptist Noel and I had much conversation about Free Church matters ; he also expressed an anxious desire that something should be done in St. Andrews to turn M. Guizot's mind to serious religion. He knows him well, and feels a great interest in his happiness. His misfortunes may, of themselves, turn his thoughts and affections heavenwards.

" *P.S.*—I met Dr. Somerville on the street the day of my arrival here, looking as young as when I saw him last. He and Mrs. Somerville were to set out next day for Kelso to visit his sister, Mrs. General Elliot."

The following short extract contains an interesting allusion to his early days :—

" ROSSIE PRIORY, *Jan.* 1849.

"Mr. Graham, the Established clergyman of Abernyte, dined here yesterday. He has a great passion for optics, and has made some very fine and large telescopes with his own hands. Although the day confined everybody to the house, I took a walk of two miles to pay him a visit, which was a very agreeable one. He has a nice clever wife, but no children, and his manse is one of the most charming residences I have seen. It was a great treat to me to find a young man carrying on all my early pursuits, and who had derived his practical knowledge from my own writings."

In 1849 he received a mark of distinction which he highly valued, being chosen one of the eight Foreign Associates of that French Institute which had done him honour so early in his career. He succeeded Berzelius, the celebrated Swedish chemist. Of this honour, the greatest scientific one which France can bestow on foreign sages, and which, alike under republic and under monarchy, has been exercised with judgment and discretion, Baron Cuvier remarked that it was one " for which all the philosophers of Europe compete, and of which the list, beginning with the names of Newton, Leibnitz, and Peter the Great, has at no period degenerated from its original lustre."

In the first month of 1850, the second dark cloud of his life overshadowed my father, not bursting like a sudden thunderstorm, as on the first occasion, but creeping slowly on during years of delicate health and prostration of strength, which, although without disease, was so great, that at almost the first touch of an epidemic the enfeebled constitution succumbed, and, after a week's illness more serious than usual, the wife of his youth passed away on the 27th of January, in a quiet humble hope through that way of simple salvation which her sorrowing husband had not yet entered with his heart, though he understood it with his intellect. She was laid to rest beside her long-loved and long-mourned " Charlie," beneath the shade of the Abbey ruins of Melrose, and within the sound of the rippling river which had caused her greatest bereavement.

## CHAPTER XIII.

### NOTES OF LIFE IN 1850-51.

They come !
My ear drinks in the measured tread,
    It steals upon me from afar ;
They come ! but not o'er heaps of dead,
    Their tread is not the tramp of war.

They come !
But no red carnage tracks their heel,
    No blood-stained banners do they wave ;
They carry not the murderous steel,
    Nor dig at each new step a grave.

They come !
Their numbers wave like ears of corn
    Before the wind ; their ranks increase ;
But theirs are numbers that adorn
    The armies of ' *The Prince of Peace.*'

They come !
(Though all unheeded comes their band,
    Its work not known, nor understood)
To knit together every land
    In one vast bond of brotherhood.

They come !
And men cry, ' Dreamers, get ye gone !
    Avaunt ! *What will these babblers say !*
What ! know ye not ye stand alone,
    With nought to prop your cause a day ?'

They come !
And answer, ' *Not by might or power*
    Shall this our Gospel cause hold good ;
But yet as sure as comes the hour
    Shall come our bond of brotherhood.'
    *      *      *

Rev. Dr. ASPINAL *of Liverpool.*

ON the 15th of April 1850 we went abroad for change
of air and scene, after the heavy pressure of desolating
bereavement,—first visiting Brussels and Antwerp, and

then remaining for some weeks in Paris, where my father enjoyed meeting scientific and literary friends, M. de la Rive, the Abbé Moigno, editor of *Les Mondes*, M. Babinet, the Chevalier Neukomm, and others; he was much with Lord Brougham, who was in Paris that spring, and came to see us almost every day. Even when he had not time to come in, his quaint friendly visage was sure to appear at a small opening of the antechamber door, from whence he poured forth original and striking conversation. But the man to whom my father's heart clung most, and who excited his warmest admiration and sympathy, was Dominique François Jean Arago, five years his junior, whose acquaintance he had first made when in the prime of manhood, at the French Institute in 1814, and of whose early and chequered career Brewster had given some notes in his home letters. Since then life had been to the French astronomer full of changes and vicissitudes. The following contemporary notices of Arago during our visit to Paris may not be uninteresting :—

" One morning, in the spring of 1850, we were musing somewhat uneasily in our apartment at the Hôtel Bristol, in the Place Vendôme. Only recently arrived from our own peaceful land, the rumours which prevailed of the certainty of an *émeute*, and the probability of another Revolution, were not particularly cheering. Lord ——, the highest diplomatic authority, had assured my father that there was no danger of an *émeute* ' before Saturday.' Many of our countrymen had provided passports, to be in readiness for instant flight; while the mounted guard at each end of the principal streets, and the frequent march past our windows of armed troops, with their magnificent military music, gave a sort of substance to the shadows of report.

" On the morning alluded to a tap at the door of the little ante-room announced a visitor; and in walked one whom it was easy to pronounce, without hearing him individually named, to be one of the few ' whose names are not born to die.' He was tall, though somewhat bent, with hair grizzled and matted, eyes deeply sunk, and lofty brow, furrowed more with sorrow, care, and labour than with age; his features were not handsome, yet was there something in them grand and massive, and expressive at once of expansive intellect and of the deepest depression. One's heart turned towards him instinctively with a tenderness and reverence, not lessened when told that he was Arago! He spoke no English, although reading and understanding it with ease; and presently a flood of scientific conversation, in a medley of French and English, threatened to sweep away the remembrance that any other character but that of a philosopher belonged to this great Frenchman. Arago, however, to the mind's eye, was present as the republican—as the Minister of War and Marine, during the short-lived Provisional Government of 1848—as the bearer of the white flag at the murderous barricades, while bravely attempting to stem the awful passions to which republicanism had given a form and a consistency. However the incongruity may have been regretted, there was something noble and disinterested in the patriotic feeling which drew the philosopher from his study—the astronomer from his quiet heavens—thus to do and dare. At length I ventured to interpose a few words,

amidst the thick-coming theories, demonstrations, and
discussions—to make anxious inquiries as to the truth
of the rumoured danger.    A cheerful laugh, and '*Bah !
bah ! Paris est assez tranquille,*' had a wonderful effect
in banishing the dreams and nightmares of an excited
imagination.    He went on to give his opinion as to the
state of the people, the false reports and exaggerations
so currently circulated and believed, and the improba-
bility of further danger and bloodshed, at least for a
time, which subsequent events fully verified.    There
was something in his voice and look that gave one confidence.

"Some days after we went to the Observatoire—a large
and magnificent building erected by Louis Quatorze,
of which M. Arago was director, and in which he had
his home of many years, and much bereavement—one,
too, well suited to the rugged grandeur of his appearance, to the abstruse nature of his pursuits, and to the
comparative loneliness and quietude of his life.    There
he dwelt, amidst the instruments and books, which were
his chief consolation for the sorrows of life.    There he
lived, with the external heavens brought close to him
by means of the magnificent telescopes of the Observatoire—those starry, beautiful heavens, which yet he
could not see.    Yes ! Arago was blind, or nearly so ;
for, besides a constitutional tendency to this malady,
the nerves of his eyes had never recovered the shocks
they encountered in that bloody Parisian summer of
1848, when the muskets and the cannon of the infu-
riated insurgents were turned against the brave peace-
maker, although he wonderfully escaped further injury.
We were shown into Arago's library—into that room
of thought, where had been forged so many mental

levers to stir up the minds of men. It was a large and
lofty chamber, hung with prints and photographs of contemporary *savans* of all nations. Books and pamphlets
were heaped on every chair, and the tables were covered
with scientific instruments. The philosopher soon came
in, habited in his *robe de chambre*, and it was easy to
read upon his expressive countenance the traces of a
new and deep depression. That morning his attached
friend and scientific companion, M. Gay-Lussac, the
celebrated chemist, had breathed his last at his dwelling
in the Jardin des Plantes, and it was touching to see
the tender feeling and deep sense of bereavement, so
rare in a man no longer young, and of such absorbing
thought and occupation. He spoke a few words, also
very despondingly, of his own health, and seemed to
anticipate that the close of his life was near at hand.
His cheerfulness, however, partially returned, as his old
friend directed the conversation to his favourite topics.
It was indeed a touching and an interesting sight to
watch the communing of those two remarkable men, so
different, yet so full of sympathies, which they never
again were to interchange in life.

" Before saying farewell to Arago, we went all through
the spacious halls and beautiful machinery of the Observatoire—all so well worthy of admiration, as well as
interesting from being the home and scene of labour
of so eminent a man. At last we emerged upon the
top, from whence we saw the most beautiful view
that could be imagined of that strangely fascinating,
wonderful city, with its spires and faubourgs, Seine and
bridges, stretched out before us as calmly and silently
as if there existed not within it so many appalling
elements of woe, crime, and anarchy. A little inci-

dent, that occurred at the door of the Observatoire, threatened to recall our *rouge* fever.    When we were seated in the carriage, one of the *savans*, who had been doing the honours of the Observatoire, had confessed his royalist tendency, and abhorrence of the much-abused words, *Liberté, Egalité, Fraternité*, which were stuck up in all directions ; in the eagerness of the conversation we forgot the dulcet tones which such a subject then necessitated in Paris, and the French philosopher was just presenting us with a small tumbler, royally ciphered, which had been rescued from the sack of the Tuileries, as a memorial of royalist sympathies, when we became aware of having attracted the attention of the French footman and coachman on the box, who were listening with an unmistakable earnestness of attention and ferocity of look.    The philosopher, with somewhat unphilosophic haste, changed the subject, and we all immediately found the weather to be a topic of engrossing interest.    We drove away from the Observatoire with a melancholy feeling that we should never again see its distinguished *chef*, which indeed was the case.    On our next visit we found that Arago was a prisoner from a severe attack of illness, and shortly afterwards we took our final departure from Paris."

Little more than two years after this, Arago lay down to die, and in connexion with that event, a conversation was recorded in scientific annals which gives at least the consolation of knowing that Arago was possessed of a praying mother and a faithful friend.    An old friend of my father's, M. de la Rive, a Swiss philosopher, whom it was a great happiness to meet in Paris at this time, had been led by the loss of his wife to seek and to find

consolation in revealed religion.  Being again in Paris
at the time of Arago's last illness, M. de la Rive sought
to render his illustrious friend a partaker of the same
happiness, although fully understanding the difficulties
of a mind which would only admit what it could perfectly comprehend.   He thus described the interview :
— " We conversed upon the marvels of creation and
the name of God was introduced.   This led Arago to
complain of the difficulties which his reason experienced
in understanding God.   ' But,' said I, ' it is still more
difficult not to comprehend God.'   He did not deny it.
' Only,' added he, ' in this case I abstain, for it is impossible for me to understand the God of you philosophers.'   ' It is not with them that we are dealing,'
replied I, ' although I believe that true philosophy
necessarily conducts us to belief in God: it is of the
God of the Christian that I wish to speak.'   ' Ah,' he
exclaimed, ' He was the God of my mother, before
whom she always experienced so much comfort in
kneeling.'   ' Doubtless,' I answered.   He said no more ;
his heart had spoken ; this time he had understood." [1]

On the 2d of October 1853 Arago breathed his last,
at the age of sixty-seven.   His last words were—" *Travaillez, travaillez bien ;*" and a new edition of three
volumes of scientific notices, it was touchingly said, " a
été préparée par Arago mourant."   The Emperor decreed
a public funeral for this man, so widely celebrated and
so deeply beloved.   In spite of a heavy rain, the procession was followed or awaited by crowds in silent and
tearful sorrow.   As many as twelve thousand persons
thus " assisted " at the great mourning, proving " that

[1] *François Arago.*   Par M. de la Rive.   *Bull. Univ.* de Genève.  Oct.
1853.

the name of Arago had preserved all its prestige and its immense popularity." In the Cemetery of Père la Chaise rest the remains of Arago, the mourner, the biographer, the republican, the statesman, the philosopher!

My father was deeply interested in all M. Arago's works, and reviewed several of them in the *North British*. In a notice of Arago's Life, he thus wrote :—" We have had the good fortune, as we now feel it, of breaking a lance with Arago, both as a principal and a second, in some of the tournaments of science. A nobler and more generous opponent we never encountered. When after a campaign of twenty-five years it became necessary that we should meet, he prepared the way by a letter of lofty sentiment and warm affection. Other twenty years have elapsed, in which we have found ourselves in open combat with him on questions of exciting interest and national feeling; but he has ever shown to us the warmest friendship, not only in words which he has addressed to the world, but in acts of substantial and much valued kindness. It is therefore with the deepest sorrow that we mourn the double loss of a friend and of a sage, and that we now express over his tomb our admiration of his genius, our sympathy with his patriotism, our gratitude for his kindness, and our affection for his character."

An evening spent with M. Guizot at this time interested my father exceedingly. It was strangely solemnizing to enter from the guarded yet threatening streets, the scene of the *coup d'état* only a year later, into the quiet salon of the fallen statesman. He showed us the portrait of his father, who had perished on a scaffold of the first Revolution, and for whom Guizot's mother had worn mourning to her last days. In that happy intelligent home-circle, however, it was difficult to retain long the remembrance of outside turbulence.

Amidst much sight-seeing in Paris and its neighbourhood, nothing occupied his attention more than the remarkable electro-chemical telegraph, exhibited to us by its inventor, Mr. Alexander Bain, upon which M. Leverrier and Dr. Lardner were experimenting at that time before committees of the Institute and the National Assembly. Brewster afterwards wrote :—
"When we saw in Paris the whole operation of perforating the message and recording it in blue lines at the other end of the wire, it seemed more like magic than any result of mechanism which we have ever seen. The dry steel point, when tracing its spiral path, actually seems to be depositing blue ink upon the paper. But it is not merely ingenuity that is the characteristic of Mr. Bain's telegraph. It is unlimited in its quickness, and unerring in its accuracy ; and it has another advantage, of requiring a battery of much less power than other forms of the telegraph." Mr. Bain received a large sum for his inventions from the Electric Telegraph Company, and he went afterwards to America, where his form of telegraph is, I believe, still extensively used, and under some conditions of the atmosphere it works better than the ordinary kind. The fact of Mr. Bain being a Scotchman did not decrease the admiration with which the beautiful experiments were watched. Although the first steps in the invention of the telegraph were taken in France by M. Le Monnier, and in England by Sir W. Watson about the year 1747, yet the idea of its practical application to the transmission of messages was

undoubtedly first suggested by an anonymous correspondent of the *Scots Magazine*, dated Renfrew, Feb.
1, 1753, signed C. M., and entitled "An Expeditious
Method of conveying Intelligence." After a good deal
of correspondence on the subject, Sir David Brewster
gave up all hope of discovering the name of the inventor,
and it was not till 1859 that he had the great pleasure
of solving the mystery, in the following manner : he
"received from Mr. Loudon of Port-Glasgow a letter,
dated 31st Oct. 1859, stating that while reading the
article in the *North British Review* his attention was
arrested by the letter of C. M., and having mentioned
the fact to Mr. Forman, a friend then living with him,
he told him that he could solve the mystery regarding
these initials.     Mr. Forman recollects distinctly of
having read a letter, dated 1750, and addressed by his
grandfather, a farmer near Stirling, to Miss Margaret
Winsgate, residing at Craigengilt, near Denny (to whom
he was subsequently married), referring to a gentleman
in Renfrew of the name of Charles Morrison, who
transmitted messages along wires by means of electricity,
and who was a native of Greenock, and bred a surgeon.
Mr. Forman also states that he was connected with the
tobacco trade in Glasgow—that he was regarded by the
people in Renfrew as a sort of wizard, and that he was
obliged, or found it convenient, to leave Renfrew and
settle in Virginia, where he died.     Mr. Forman also
recollects of reading a letter in the handwriting of
Charles Morrison, addressed to Mr. Forman, his grandfather, and dated 25th Sept. 1752, giving an account of
his experiments, and stating that he had sent an account
of them to Sir Hans Sloane, the President of the Royal
Society of London, who had encouraged him to perfect

his experiments, and to whom he had promised to publish an account of what he had done.   In this letter
Mr. Morrison stated that as he was likely to be ridiculed
by many of his acquaintances, he would publish his
paper in the *Scots Magazine* only with his initials."

After a week or two in London, where we made the
acquaintance of Frederica Bremer, whose kindly, homely
simplicity my father much admired, we came northward,
paying a visit *en route* to Mr. and Mrs. Fox Talbot in
their lovely summer residence on the banks of Winder-
mere, where the two philosophers had much pleasant
intercourse over photography and other scientific pursuits, rendered doubly interesting by the beauty of the
scenery.

The twentieth meeting of the British Association took
place for the second time in Edinburgh, July 23d of this
year, and Sir David Brewster was the President.   His
opening address was carefully prepared, as all his public
appearances were, and was published separately.   True
to his persistent energy of aim, whatever others might
do, he never forgot the original objects of this Association, which he again brought forward prominently.   He
spoke as follows :—" It has always been one of the leading objects of the British Association, and it is now the
only one of them which has not been wholly accomplished, ' to obtain a more general attention to the objects
of science, and a removal of any disadvantages of a public
kind which impede its progress.'   Although this object
is not very definitely expressed, yet Mr. Harcourt, in
moving its adoption, included under it the revision of
the law of patents, and the direct national encouragement
of science,—two subjects to which I shall briefly direct
your attention.   In 1831, when the Association com-

menced its labours, the patent-laws were a blot on the legislation of Great Britain; and though some of their more obnoxious provisions have since that time been modified or removed, they are a blot still, less deep in its dye, but equally a stain upon the character of the nation. The protection which is given by statute to every other property in literature and the fine arts, is not accorded to property in scientific inventions and discoveries. A man of genius completes an invention, and, after incurring great expense, and spending years of anxiety and labour, he is ready to give the benefit of it to the public. Perhaps it is an invention to save life —the life-boat; to shorten space and lengthen time— the railway; to guide the commerce of the world through the trackless ocean—the mariner's compass; to extend the industry, increase the power, and fill the coffers of the State—the steam-engine; to civilize our species, to raise it from the depths of ignorance and crime to knowledge and to virtue—the printing-press. But, whatever it may be, a grateful country has granted to the inventor the sole benefit of its use for fourteen years. That which the statute freely gives, however, law and custom as freely take away, or render void. Fees, varying from £200 to £500, are demanded from the inventor; and the gift, thus so highly estimated by the giver, bears the great seal of England. The inventor must now describe his invention with legal precision. If he errs in the slightest point—if his description is not sufficiently intelligible—if the smallest portion of his invention has been used before—or if he has incautiously allowed his secret to be made known to two individuals, or even to one—his patent will be invaded by remorseless pirates, who are ever on the

watch for insecure inventions, and he will be driven
into a court of law, where an adverse decision will be
the ruin of his family and his fortunes. Impoverished
by official exactions, or ruined by legal costs, the hapless
inventor, if he escapes the asylum or the workhouse, is
obliged to seek, in some foreign land, the just reward of
his industry and genius. Should a patent escape un-
scathed from the fiery ordeal through which it has to
pass, it often happens that the patentee has not been
remunerated during the fourteen years of his term. In
this case, the State is willing to extend his right for five
or seven years more ; but he can obtain this extension
only by the expensive and uncertain process of an Act
of Parliament—a boon which is seldom asked, and
which, through rival influence, has often been with-
held. Such was the patent-law twenty years ago ; but
since that time it has received some important amelio-
rations ; and though the British Association did not
interfere as a body, yet some of its members applied
energetically on the subject to some of the more
influential individuals in Lord Grey's Government, and
the result of this was, two Acts of Parliament, passed in
1835 and 1839, entitled, ' Acts for Amending the Law
touching Letters Patent for Inventions.' . . .

"The other object contemplated by the British
Association—the organization of science as a national
institution—is one of a higher order, and not limited to
individual or even to English interests. It concerns
the civilized world :—not confined to time, it concerns
eternity. While the tongue of the Almighty, as Kepler
expresses it, is speaking to us in His Word, His finger
is writing to us in His works ; and to acquire a knowledge of these works is an essential portion of the great

o

duty of man. Truth secular cannot be separated from
truth divine; and if a priesthood has in all ages been
ordained to teach and exemplify the one, and to maintain, in ages of darkness and corruption, the vestal fire
upon the sacred altar, shall not an intellectual priesthood
be organized to develop the glorious truths which time
and space embosom—to cast the glance of reason into
the dark interior of our globe, teeming with what was
once life—to make the dull eye of man sensitive to the
planet which twinkles from afar, as well as to the
luminary which shines from above—and to incorporate
with our inner life those wonders of the external world
which appeal with equal power to the affections and to
the reason of immortal natures?  If the God of Love is
most appropriately worshipped in the Christian Temple,
the God of Nature may be equally honoured in the
Temple of Science.  Even from its lofty minarets the
philosopher may summon the faithful to prayer; and
the priest and the sage may exchange altars without the
compromise of faith or of knowledge."

The subject of the patent-laws had long occupied
the attention of Brewster, and Lord Brougham was the
champion of this long-contested battle.  In 1835 Lord
Brougham's first bill to give relief to the sorely op-
pressed inventors of England was passed by a considerable majority in the House of Lords, but only by
a small majority in the House of Commons.  Much
was still needed, and after long efforts the amalgamated
bill of Lord Brougham and Lord Granville was passed
in 1852.  So many alterations were made in Committee that the Act was still regarded only as a mere
instalment of reform.

In 1865 Sir David Brewster wrote upon this subject:

—"The injustice of the patent-law has been so fully admitted, that various Acts of Parliament have been passed in favour of the patentee, adding slightly to the protection of his right, and reducing the expense of its attainment; but no addition has been made to the shortness of its tenure, and no increase of security against direct piracy, or partial infringement. Whatever difficulty the statesman may experience in giving security to the rights of inventors, he can have none in giving them the same tenure as copyrights, and conferring them as gratuitously, or at no greater cost than is necessary to cover the expenses of the Patent Office. Between the national claims of authors and inventors there can be no comparison. Value as you may, and value highly, the treasures of ancient and of modern thought, what are they when weighed against the inventions of art and science, predominating over our household arrangements, animating our cities with the sounds of industry, and covering with mechanical life the earth and the ocean? The eloquence of the orator, the lesson of the historian, the lay of the poet, are, as it were, but the fragrance of the plant whose fruit feeds us, and by whose leaves we are healed; or as the auroral tint which gives a temporary glory to a rising or a setting sun. But grant to the favoured genius of copyright its highest claims, and appreciate loyally its most fascinating stores, their value is shared, and largely shared, with that of the type, the paper, and the press, by which these stores have been multiplied and preserved. The relative value of books and inventions may be presented under another phase. Withdraw from circulation the secular productions of the press that are hoarded in all the libraries of the world, and

society will hardly suffer from the change.    Withdraw
the gifts with which art and science have enriched us
—the substantial realities through which we live, and
move, and enjoy our being—and society collapses into
barbarism."

In 1851 the Duke of Argyll was elected Chancellor
of the University of St. Andrews, and his installation
took place in the large Hall of the University Library,
—a ceremony which many crowded to witness.    The
Duke and Duchess of Argyll, and Lady Emma Campbell,
visited my father on this occasion at St. Leonard's
College.

This was a busy year to him, as to many of his com-
peers.    His duties as a juror of the Great Exhibition
kept him in London for some months.    During this
time the Kohinoor diamond was an object of great
interest to him.    The various phenomena to be observed
in precious stones, their fluids, their embedded crystals,
and their " pressure cavities " had long been traced and
experimented upon by him.    He had invented an instrument for testing and examining precious stones, which
he called a Lithoscope, from two Greek words signifying *a stone*, and *to see*—one constructed by Dollond
having been exhibited at the British Association at
York in 1832.    Ladies brought their jewels to him to
be admired and examined, and were startled to receive
them branded as simple glass.    As long ago as the old
Kinrara days a splendid set of amethysts shown to him
by Jane Duchess of Gordon, of witty memory, were
ruthlessly denounced as " shams."

The diamond and its strange history—brilliant production as it is of dim and dark vegetable life—had
been a special object of research to him.    In compara-

tively early life Brewster was the first person to investigate the remarkable optical structure of the diamond ;
and an early scientific friend of his, Sir George Mackenzie of Coul, was the first person in this country who
burned diamonds, making a free use of his mother's
jewels, and by means of diamond powder converting
iron into steel.	Sir David was therefore quite in his
element while examining the famous gem which drew
so many admirers around it in the Crystal Palace.	At
first it had caused disappointment, as owing to its
position and the manner in which it was cut, it emitted
little brilliancy, but when, at his suggestion, fifteen or
sixteen gas lights were placed behind, it threw out a
radiance of coloured light which delighted all who saw
it.	In 1852, having been consulted along with others
by Prince Albert as to the best manner of having it
re-cut, he was kindly given every facility of examining
it at Buckingham Palace, which he did with the micro-
scope and by the aid of polarized light.	This further
minute investigation only confirmed the conclusion he
had previously arrived at, that this diamond, large and
beautiful as it was, was not the Mountain of Light, nor
any portion cut from the original body, although given
to the English under that name from the Lahore Treasury, where it had been placed after the death of
Runjeet Sing.

He wrote as follows :—

" LONDON, *May* 31, 1851.

" . . . I was occupied all yesterday in examining
the diamonds in the British Museum.	The Duke of
Northumberland had told me in the forenoon that it
was a general belief in India that the Kohinoor dia--
mond in the Queen's possession is not the real one which

belonged to the Great Mogul, and which was weighed
and examined by Tavernier in 1665, and I went to the
library of the Athenæum to see Tavernier's drawing and
description of it. From both it is obvious that the
Queen's diamond is not the Kohinoor—The MOUNTAIN
OF LIGHT. I send a sketch of Tavernier's drawing of it,
which is like a *mountain*, resembling, as he says, half an
egg, or one cut in two, whereas the Queen's diamond
has no such form, and could not be obtained from the
above by any process of cutting. Besides, the above
weighed 280 carats, and the Queen's only 184 carats.
The real Kohinoor of the Great Mogul, when in a rough
state, weighed 787½ carats, and was reduced by a Vene-
tian diamond cutter, the Sieur Hortensio Borgis, to 280,
which so enraged Shah Jehan that he refused to pay
him for his labours, and made him pay as a fine 10,000
rupees."

In all my father's writings one thing is prominent,
and that is the care he ever took to draw a religious
moral from his subject-matter; thus he concludes an
article on the Diamond in these words:—"A moral
as well as a secular lesson is read to us by the diamond. Like every organism of this world, it bears
the impress of decay. The stoutest metal and the
toughest gem exist by forces which time weakens and
the elements destroy; and in that great catastrophe
when the 'earth and the works which are therein, shall
be burned up,' the jewel so highly prized will pass into
its primeval cinder, while the silver and the gold will
only change their form and reappear perchance brighter
and purer in the new earth which is to arise. Let us
covet, then, the virgin gold and the pure silver of truth

and justice, and estimate at their real value the glitter-
ing qualities and the dazzling possessions which bear
so high a value in this world, but which have none in
the next."

My father's horror of war and its appalling train of
consequences, which he considered as a scene of legalized
slaughter, was a very marked feature of his mind. He
considered it alike a breach of the commandment of
Sinai and of the spirit of the New Testament, and to the
last, never softened the statement which he so frequently
made through life, that "he could not understand how
any Christian could be a soldier." That he had two
sons in the army was a real grief to him, to which he
alludes in an address from which I shall presently quote.
Nothing excited his indignation more than any encouragement in the pulpit or from clerical lips of the
miserable glories of martial fame. His attention had
been much directed to the subject of international peace,
by means of international arbitration. It was in accordance, therefore, with all his views and feelings, that
when requested to act as President of the great Peace
Congress, held in the Exhibition year, he should put
aside his dislike to such a prominent position, and
lend his energies to perform it thoroughly. He thus
wrote :—

" LONDON, *July* 22, 1851.

" I HAVE just come from Exeter Hall, where I have
been presiding as President of the Congress of Peace,
which meets for three days, the 22d, 23d, and 24th.
On Friday last a deputation from the Congress, consisting of Mr. Cobden and other two gentlemen, called upon
me to ask me to be the President of the Congress at its
meetings in London. I, of course, refused on account

of my incompetency as a speaker; but having learned
that it was not a speech that I had to make, but an
address, which was read by the Presidents of the three
last meetings at Paris, Brussels, and Frankfort, I agreed,
as I had just sent off the last page of my article to
Edinburgh. My engagements were so numerous that
I could scarcely find time to write my address. I,
however, set myself heartily to the task, and I, this day,
delivered it with much courage, to a splendid audience
of nearly 4000 persons, by whom it was well received.
There were several splendid speeches delivered to-day
at the Hall by Mr. Angell James, Rev. Mr. Brock,
Don Cubii Soler, a Spaniard; Mr. Athanasius Coquerel,
of Paris, a clergyman; M. Vischers of Brussels; Dr.
Beckwith, an American, and Dr. John Burnet, a Scotchman. The British members of the Congress give a
soirée to the foreign members on Friday evening, in
Willis's Rooms.

" To-morrow is the last day of our jury labours, so far
as the great medals are concerned, so that I can only be
*an hour* in the chair of the Peace Congress, where there
are to be some eloquent speakers. A large body of
French workmen of the superior class are to be there,
to testify their hatred of war. They are, of course,
visitors to the Exhibition."

This address, prepared under difficulties, possesses a
peculiar interest, from the fact of his singular appreciation of it, which was very uncommon with regard to his
own writings. Lady Brewster tells me that when, in
after years, she spoke to him of any of his compositions,
he used to say,—" Oh, they are nothing, but the Peace
address *is* worth reading." I give some extracts from it :—

" The question, 'What is war?' has been more frequently asked than answered; and I hope that there
may be in this assembly some eloquent individual who
has seen it in its realities, and who is willing to tell us
what he has seen. Most of you, like myself, know it
only in poetry and romance. We have wept over the
epics and the ballads which celebrate the tragedies of
war. We have followed the warrior in his career of
glory without tracing the line of blood along which he
has marched. We have worshipped the demigod in the
Temple of Fame, in ignorance of the cruelties and crimes
by which he climbed its steep. It is only from the
soldier himself, and in the language of the eye that has
seen its agonies, and of the ear that has heard its
shrieks, that we can obtain a correct idea of the miseries
of war. Though far from our happy shores, many of us
may have seen it in its ravages and in its results, in
the green mound which marks the recent battle-field,
in the shattered forest, in the razed and desolate village,
and, perchance, in the widows and orphans which it
made! And yet this is but the memory of war—the
faint shadow of its dread realities—the reflection but of
its blood, and the echoes but of its thunders. I shudder
when imagination carries me to the sanguinary field, to
the death-struggles between men who are husbands and
fathers, to the horrors of the siege and the sack, to
the deeds of rapine, violence, and murder, in which
neither age nor sex is spared. In acts like these the
soldier is converted into a fiend, and his humanity even
disappears under the ferocious mask of the demon or
the brute. To men who reason, and who feel while
they reason, nothing in the history of their species
appears more inexplicable than that war, the child of

barbarism, should exist in an age enlightened and
civilized, when the arts of peace have attained the
highest perfection, and when science has brought into
personal communion nations the most distant, and races
the most unfriendly. But it is more inexplicable still
that war should exist where Christianity has for nearly
2000 years been shedding its gentle light, and that it
should be defended by arguments drawn from the
Scriptures themselves. When the pillar of fire con-
ducted the Israelites to their promised home, their Divine
Leader no more justified war than he justified murder
by giving skill to the artist who forges the stiletto, or
nerve to the arm that wields it. If the sure word of
prophecy has told us that the time must come when
men shall learn the art of war no more, it is doubtless
our duty, and it shall be our work, to hasten its fulfilment, and upon the anvil of Christian truth, and with
the brawny arm of indignant reason, to beat the sword
into the ploughshare, and the spear into the pruning
hook. I am ashamed, in a Christian community, to
defend on Christian principles the cause of universal
peace. He who proclaimed peace on earth and good-
will to man, who commands us to love our enemies, and
to do good to them who despitefully use us and per-
secute us ; He who counsels us to hold up the left cheek
when the right is smitten, will never acknowledge as
disciples, or admit into His immortal family, the sovereign
or the minister who shall send the fiery cross over
tranquil Europe, and summon the bloodhounds of war
to settle the disputes and gratify the animosities of
nations. The cause of peace has made, and is making,
rapid progress. The most distinguished men of all
nations are lending it their aid. The illustrious Hum-

boldt, the chief of the republic of letters, whom I am proud to call my friend, has addressed to the Congress of Frankfort a letter of sympathy and adhesion. He tells us that our institution is a step in the life of nations, and that, under the protection of a superior power, it will at length find its consummation. He recalls to us the noble expression of a statesman long departed, ' that the idea of humanity is becoming more and more prominent, and is everywhere proclaiming its animating power.' Other glorious names sanction our cause. Several French statesmen, and many of the most distinguished members of the Institute, have joined our alliance. The Catholic and the Protestant clergy of Paris are animated in the sacred cause, and the most illustrious of its poets have brought to us the willing tribute of their genius. Since I entered this assembly, I have received from France an olive branch, the symbol of peace, with a request that I should wear it on this occasion. It has lost, unfortunately, its perishable verdure—an indication, I trust, of its perennial existence."

After an eloquent description of the crystal "Temple of Peace," which was drawing crowds of different sea-severed nationalities, he goes on :—

" Amid these proud efforts of living genius, these brilliant fabrics, these wondrous mechanisms, we meet the sage, and the artist of every clime and of every faith, studying the productions of each other's country, admiring each other's genius, and learning the lessons of love and charity which a community of race and of destiny cannot fail to teach. The grand truth, indeed, which this lesson involves, is recorded in bronze on the prize medal by which the genius of the exhibitors is to be rewarded. Round the head of Prince Albert, to whose talent and moral courage we owe the Exposition of 1851, and addressed to us in his name, is the noble sentiment : 'Dissociata in locis concordi Pace ligavi'—'What space has separated I have united in harmonious peace.' This is to be our motto, and to realize it is to be our work. It will, indeed, be the noblest result of the Prince's labours, if they shall effect among nations what they have already done among individuals, the removal of jealousies that are temporary, and the establishment of friendships that are enduring. Nations are composed of individuals, and that kindness and humanity which adorn the single heart, cannot be real if they disappear in the united sentiment of nations."

This was the fourth meeting of the Peace Congress, which had been originated by Mr. Elihu Burritt. On the first day letters of adhesion were read from Count Pierre Dionysie Dumelli, President of the Chamber of Deputies of Turin, and from Thomas Carlyle, a friend and correspondent of the President for many years, and who in this matter heartily sympathized with him. These were his characteristic words :—

" I fear I shall not be able to attend any of your meetings ; but certainly I can at once avow—if indeed such an avowal on the part of any sound-minded man be not a superfluous one—that I altogether approve your object, heartily wish it entire success, and even hold myself bound to do, by all opportunities that are open to me, whatever I can towards forwarding the same. How otherwise ? 'If it be possible, as much as in you lies, study to live at peace with all men ;' this, sure enough, is the perpetual law for every man, both in his individual and his social capacity ; nor in any capacity or character whatsoever is he permitted to neglect this law, but must follow it, and do what he can to see it followed. Clearly, beyond question, whatsoever be our theories about human nature and its capabilities and outcomes, the *less* war and cutting of throats we have among us, it will be the better for us all!"

The second day very interesting letters of adhesion were read from "several distinguished foreigners, amongst whom were M. Barthélemy St. Hilaire, Member of the National Institute, representative of the people, and formerly ambassador to England; M. Carnot, representative of the people, and son of the celebrated Carnot who organized Napoleon's armies; M. Victor de Tracy, formerly Minister of Marine in the administration of M. Odillon Barrot; Dr. Bodenstedt, and General Subervie, one of the oldest generals in France. The latter wrote : 'Never was there a cause more holy; of all the scourges that can afflict the world, war is the most terrible.' "

The third day letters of sympathy and approval were read from Archbishop Whately, M. Victor Hugo, and, on one of the days, from the Archbishop of Paris. The President concluded the proceedings of the Congress by one of those practical hints which he always strove to introduce :—

" Were our youth better instructed than they are in the popular departments of physical and natural science, subjects with which no deeds of heroism or personal adventure are associated ; and were every school to have a museum containing objects of natural history, and specimens of the fine and the useful arts, the amusements of the school would assume a different character, and the scholars would go into active life better fitted for those peaceful professions to which ere long they must be confined.    But there is still another class whose active interest in the cause of peace I would fain secure. If there are mothers in this assembly, as I can testify that there are fathers, whose sons have been sent in the service of their country to the regions of pestilence or of war, I need not solicit their assistance in propagating the doctrines of peace.    They will proffer it in tears— in tears shed in the recollection of those anxious days in which they have followed in their hazardous career the objects of their deepest love,—now sinking under a burning sun, now prostrate under tropical disease, now exposed to the sword of the enemy."

During this summer, Sir David, notwithstanding his busy labours as a jurist, found time as usual for much society, and thus wrote :—

" I dined with Mr. Cowan, M.P., on the 16th, and with Sir Robert Inglis and a nice party on the 17th, and went in the evening to Lord Rosse's second soirée, bristling with foreigners.    On the 19th, on coming home from the Exhibition to dress, I was surprised by an invitation to the Queen's ball, to which I of course went. It was a splendid sight, and I met there with crowds of friends.    The Queen danced a great deal, and there was something in her whole manner (so happy and cheerful and frank), and in that of the Prince, which made the most favourable impression on everybody.    The apart- ments in the Palace were all thrown open, and the party was very numerous.    There were refreshments— tea, coffee, ices, etc., in one room, and a standing supper in the dining-room.    We got home about three in the morning, after waiting about an hour in the lobby, where some ladies were sleeping on their seats, and others stretched on the stone steps waiting for their carriages. We were at this time with the Herschels, and had much amusement from the scene around us. The whole display surpassed in beauty and grandeur anything I had seen.

" On the 20th I dined with Sir John Herschel, and on the 21st I had *three* parties to encounter,—one being a very agreeable dinner at the Bishop of Durham's, where Mrs. Opie was, upwards of eighty, but full of life, whose acquaintance I made. . . .

" One of the most interesting acquaintances I have made since I came here, I made yesterday. It was that of Miss Brontë, the authoress of *Jane Eyre* and *Shirley*, a little, pleasing-looking woman of about forty, modest and agreeable. I went through the Exhibition with her yesterday."

Later in the year we saw together many interesting literary personages—numberless celebrities at the Countess of Lovelace's (Ada Byron), and elsewhere ;— a pleasant breakfast at the venerable Rogers', where were Sir Charles Lyell and Dean Milman,—the poet himself being the object of the deepest interest, as, with the admiration and reverence for Scripture which distinguished him, he repeated, in tones tremulous with age and feeling, what he called the " Child's Psalm," " The Lord is my Shepherd, I shall not want." This love of the Bible in Rogers my father recollected with interest, and alluded to on another occasion :—" Since writing the above I have received a note from Lady B—, a great friend of Mr. Rogers. She says,—' I went to dear Mr. Rogers' for an hour on Tuesday evening last, and read to him a chapter of Isaiah, the 40th, and the 15th of 1 Corinthians, which pleased the dear old man, and delighted myself.'"

One of the sights which my father most keenly enjoyed was a walk by gaslight through the fairy Palace with Mrs. Davenport,[1] and Sir Charles Fox as our cicerone. It was an unusual privilege, and we were carefully watched by policemen gliding around with feline footsteps, being shod in India-rubber. The crystal roof flashed back the light till it appeared like the firmament, bright with huge planets. The sculpture gleamed or frowned in the bright light and thick shadows in which we alternately moved, till it seemed instinct with life and movement, reminding us of a simple experiment which my father often exhibited at home, moving a candle slowly round and round the face and head of a statue in an otherwise dark room, by the light and shadow of which the speaking expression of a life-like face is obtained.

Nothing could be more interesting and improving than going the round of the various departments of that wonderful Exhibition with him. His love and gift for popularizing knowledge never shone to greater advantage. He particularly enjoyed little popular scientific *séances* with groups of intelligent ladies, a pleasant custom remembered by many, and playfully alluded to in the following letter :—

"MY DEAR MRS. DAVENPORT,—I thank you very much for the privilege you have obtained for me of becoming acquainted with the Archbishop of Canterbury.

[1] Now the Dowager-Lady Hatherton.

"It will give me great pleasure to meet you on Saturday at three o'clock at the Crystal Fountain, which will not be so crowded on that day as it has been during the week.

"I shall endeavour to get up a course of lectures for you on the Paranapthadipine and all the other products of peat; but you must understand that if you come to the first lecture you must attend the whole course, and thus give me the pleasure of seeing you every day, for I mean to lecture till the close of the Exhibition. You will be glad to hear that Mr. Babbage's book has reached a second edition, which he is now busy preparing.— I am, my dear Mrs. Davenport, ever most truly yours,

D. BREWSTER."

"1 Dorset Street,
Manchester Square."

After the closing of the Great Exhibition, October 15th, we paid interesting visits *en route* homewards to Arbury Park, the beautiful residence of C. N. Newdegate, Esq., M.P., and to Capesthorne, the residence of Mrs. Davenport, who arranged an excursion to the Conway and Britannia Tubular Bridges, staying a night at Chester, and returning the next evening to Capesthorne.

This was my father's first visit to the Tubular Bridges, and possessed a peculiar interest to him from his cordial friendship with Mr. Fairbairn, and the keenness with which he had entered into the controversy as to his share of merit in the invention and construction of those marvellous monuments of engineering science. Misled by printed documents, he had, in an article in the *North British*, on the railway system, ascribed the entire credit of these works to Mr. Robert Stephenson. When put into possession of the whole facts of the case,

P

Brewster with characteristic energy defended his friend bravely in a later article solely on the subject, written after this visit, advocating what is now universally admitted,—that while Mr. Stephenson had the undoubted priority of proposing *a* Tubular bridge to span the mighty waters, *the* Tubular bridge actually doing so, owes its existence and its success to "the genius, practical knowledge, and patient experimental inquiries of the eminent engineer Mr. William Fairbairn of Manchester," who acted as engineer of the bridge, in conjunction with Mr. Stephenson, the sole engineer of the railway works.

Brewster's public appearances of this busy year were closed by an address delivered to the members of the Edinburgh Philosophical Institution on the 11th of November. I extract from it some of his geological views as brought to bear on higher science :—

"It is impossible that the human race could have existed while the world was in a state of preparation. Man could not have lived amid the storms, earthquakes, and eruptions of a world in the act of formation. The home of the child of civilisation was not ready for his reception. The stones that were to build and roof it had not quitted their native beds. The coal that was to light and heat it was either green in the forest, or blackening in the storehouse of the deep. The iron that was to defend him from external violence lay buried in the ground ; and the rich materials of civilisation—the gold, the silver, and the iron—even if they were ready, had not been cast within his reach from the hollow of the Creator's hand. But if man could have existed amid catastrophes so tremendous, and privations so severe, his presence was not required ; for his intellectual powers

could have had no suitable employment. Creation was
the field on which his industry was to be exercised, and
his genius unfolded; and that divine reason which was
to analyse and combine, would have sunk into sloth before
the elements of matter were let loose from their prison-
house, and Nature had cast them in her mould.   But
though there was no specific time in this vast chrono-
logy which we could fix as appropriate for the appearance of man, yet we now perceive that he entered with
dignity at its close.   When the sea was gathered into
one place, and the dry land appeared, a secure footing
was provided for our race.   When the waters above the
firmament were separated from the waters below it; and
when the light which ruled the day, and the light which
ruled the night, were displayed in the azure sky, man
could look upward into the infinite in *space*, as he looked
downward into the infinite in *time*.   When the living
creature after his kind appeared in the fields, and the
seed-bearing herb covered the earth, human genius was
enabled to estimate the power, and wisdom, and bounty
of its Author; and human labour received and accepted
its commission, when it was declared from on high, that
seed time and harvest should never cease upon the
earth. . . . Thus ennobled in its character, the natural
theology of animal remains appeals forcibly to the
mind, even when we regard them only as insulated
structures dislodged from the interior of the earth;
but when we view them in reference to the physical
history of the globe, and consider them as the individual
beings of that series of creations which the Almighty
has successively extinguished, and successively renewed,
they acquire an importance above that of all other
objects of secular inquiry.   The celestial creations, im-

posing though they be in magnitude, do not equal them in interest. It is only with *Life* and its associations ;— with *Life* that has been, and with *Life* that is to be, that human sympathies are indissolubly enchained. It is beside the grave alone, or when bending over its victims, that man thinks wisely, and feels righteously. When ranging therefore among the cemeteries of primeval death, the extinction and the renewal of life are continually pressed upon his notice. Among the prostrate relics of a once breathing world, he reads the lesson of his own mortality ; and in the new forms of being which have marked the commencement of each succeeding cycle, he recognises the life-giving hand by which the elements of his own mouldered frame are to be purified and re-combined."

## CHAPTER XIV.

### NOTES OF LIFE IN 1852-53.

But not alone our memories of the day ;—
Poised in mid air to greet the moonlight's ray—
Above—the jewelled crystal of the skies,
Below—the yawning gulf whence giant arms arise,—
Together we have stood ;—the midnight breeze,
The river's ripple, and the harping trees
Rejoiced themselves around.   Then upward turned
Our gaze to the far worlds that flashed and burned ;
The tower-like tube is pointed to the skies,
From the faint nebulæ new worlds arise,
The starry footsteps of the milky way
Draw swiftly near us on the giant ray,
While lunar steeps and outlined valleys shone
In the new radiance like the sculptured stone.
    Beside us, on the solemn platform now,
One stands with cheerful smile and placid brow ;
'T is not the coronet and ermined robe
That sends his honoured name around the globe—
'T is he—the giant architect who soared
To realms on high and strange new radiance poured.
    But *there* he learns the littleness of man,
The mightiness of Heaven's mysterious plan,
Permitted thus in nearness to descry
Th' illuminated oriels of the sky.

THE year 1852 opened with varied scenes.   My father
attended the marriage of his youngest son on the 2d of
January,[1] and on the 6th received the tidings of the death
of his eldest born at Mussourie, who left a widow and
two little girls.   The pressure of this bereavement was
very great, causing, as grief generally did, an increase
of the nervous irritability of his temperament.   Yet
there was evidently, though not without a struggle, a
recognition of His hand " whose trials blessed my way-

[1] To Anne Catherine, second daughter of J. C. H. Inglis, Esq. of Cramond.

ward lot." It seems to have been about this time that a MS. of an epitaph on an English Jacobite by Lord Macaulay was given to him, which was carefully preserved and much admired, as his sympathies, like those of many in Scotland, ran counter to his principles when the old Jacobite romances were in question. He was dissatisfied, however, with the want of Christian resignation expressed in it, and he therefore wrote an imitation of it, endeavouring to rectify this fault. I do not know whether the original epitaph is in print or not, but I give it now, as the two productions are interesting to compare :—

### EPITAPH ON AN ENGLISH JACOBITE.

#### By Lord Macaulay.

" To my true king I offered without stain
Courage and faith,—vain faith and courage vain ;
For him I gave land, honours, wealth away,
And one sweet hope that was more prized than they ;
For him I languished in a foreign clime,
Grey-haired with sorrow in my manhood's prime ;
Heard in Laverno, Scargill's whispering breeze,
And pined by Arno for my lovelier Tees,
Beheld each night my home in fever'd sleep,
Each morning started from the dream to weep ;
Till God, who sore me tried,—too sorely—gave
The resting-place I craved—an early grave.
Oh thou whom chance leads to this nameless stone,
From that dear country which was once my own,
By those white cliffs I never more must see,
By that proud language which I spake like thee,
Forget all feuds, and shed one English tear
On English dust,—a broken heart lies here."

### EPITAPH ON A SCOTCH JACOBITE.

#### By Sir David Brewster.

" To Scotland's king I knelt in homage true,
My heart—my all I gave—my sword I drew,

For him I trod Culloden's bloody plain,
And lost the name of father 'mong its slain.
Chased from my hearth I reach'd a foreign shore,
My native mountains to behold no more—
No more to listen to Tweed's silver stream—
No more among its glades to love and dream,
Save when in sleep the restless spirit roams
Where Melrose crumbles, and where Gala foams
To that bright fane where plighted vows were paid,
Or that dark aisle where all I loved was laid ;
And yet methought I 've heard 'neath Terni's walls,
The fever'd pulse of Foyers' wilder falls,
Or seen in Tiber's wave my Leader flow,
And heard the southern breeze from Eildon blow.
Childless and widowed—on Albano's shore
I roamed an exile, till life's dream was o'er—
Till God, whose trials blessed my wayward lot,
Gave me the rest—the early grave—I sought :
Showed me, o'er death's dark vale, the strifeless shore,
With wife, and child, and king to part no more.
O patriot wanderer, mark this ivyed stone,
Learn from its story what may be thine own :
Should tyrants chase thee from thy hills of blue,
And sever all the ties to nature true,
The broken heart may heal in life's last hour,
When hope shall still its throbs, and faith exert her power."

He afterwards went to London, whence he wrote as follows :—

"166 PICCADILLY, *May* 16, 1852.

"Although I think I do not owe you a letter, I sit down to tell you of my movements. I am just waiting for Lyon Playfair to go to an·interview with Prince Albert at three o'clock, on the subject of his great scheme of Industrial Education, upon which I am to write an article next *North British Review.* When at the Palace yesterday, examining the diamond, I lunched with the Lords in Waiting, who have a separate table, viz., Lord Charles Fitzroy, Colonel Gordon, Lord Polwarth, Colonel Phipps, etc., of the Prince's household. It was very easy and agreeable. . . .

" I have just returned from an hour-and-a-half's interview with the Prince, who unfolded to me his plan of a great central Industrial Institution, to which the £500,000 obtained from the Exhibition is to be devoted. I have been much impressed with his sagacity and knowledge and great frankness.　He told me of a letter which the Queen received from some Indian grandee, addressed to the Right Honble. Sir George Victoria, Queen of the East India Company !

" Sir James Gordon has had a fall which has injured his only knee, but he is getting better.　I hope to be able to call at the Hospital."

That this noble idea of a great centre of science and industry was perforce abandoned, greatly disappointed Brewster, who had long hoped and striven for kindred designs.　In 1865 he still was hoping, and proclaiming his hopes, although that far-seeing mind had passed away which had alone grasped the advantage of such a scheme as the proper result of the grand Industrial Exhibition of 1851.　He wrote a pamphlet entitled *Scientific Education in our Schools*, from which the following is extracted :—

" With such means in our power, cheaply obtained, and easily applied, a large portion of scientific instruction would be instilled into the youth of our schools,—familiarizing them with the works of their Maker, and preparing them for the reception of that higher revelation with which the truths of science cannot fail to harmonize. The knowledge thus imparted will not be confined to the schoolroom.　It will elevate the amusements of the holiday and the leisure hour.　It will pass into the cottage, amusing and enlightening its inmates.　It will

find its way into the workshop, giving skill to the workman, and value to his work.  It will insinuate itself
into the servants' hall, and even into the boudoir and
the drawing-room, returning an usurious interest upon
the liberality which introduced it into the school.
Thus, diffused among our now popular constituencies,
and appreciated by those above them, the truths of
science may rise into the regions of legislation, wresting
from the still reluctant statesman a measure of secular,
scientific, and compulsory education, by which the
benighted and criminal population around us may be
taught to fear God and honour the King."

In August we went to Ireland, visiting *en route* the
Duke of Argyll at Inveraray Castle, Lord Murray at
Strachur, and Mr. and Mrs. George Forbes at Rothesay.
The month spent in the "green Isle" was often recurred
to with pleasure and interest by my father, as containing very varied aspects of Irish society and character,
although it was clouded by a low feverish sort of influenza, which was the precursor of many such prostrating attacks.  Two visits to Clandeboye, the seat
of Lord Dufferin,—the one in going south, the other on
returning,—he found particularly interesting, enjoying
the combination of hereditary wit, exquisite music, and
Irish patriotism of the better sort.  A week at Birr
Castle, Parsonstown, was almost an era in the life of
the early telescope maker.  It was a touching sight
to see his delight and interest in that huge erection
of genius, the six-feet mirror Telescope, and his affinities and sympathies with its architect, Lord Rosse,
now, alas! called away from his labours.  He sat up
for several nights, as the weather, threatening at first,
cleared up sufficiently to admit of his obtaining many

interesting observations, which he thus describes in his *Life of Sir Isaac Newton*:—" We have enjoyed the great privilege of seeing and using this noble instrument, one of the most wonderful combinations of art and science which the world has yet seen. We have in the morning walked, again and again, and ever with new delight, along its mystic tube, and at midnight, with its distinguished architect, pondered over the marvellous sights which it discloses,— the satellites and belts and rings of Saturn; the old and new ring, which is advancing with its crest of waters to the body of the planet; the rocks and mountains and valleys, and extinct volcanoes of the moon; the crescent of Venus, with its mountainous outline; the systems of double and triple stars; the nebulæ, and starry clusters of every variety of shape; and those spiral nebular formations which baffle human comprehension, and constitute the greatest achievement in modern discovery." In the day-time he was always hovering around the wonderful instrument, examining its beauties and wondering at its perfections, and entering and walking through the mighty structure, which, with its huge black mouth, strongly resembled some mighty creature of a former age endowed with instinct beyond its centuries. He never tired of examining the works where the great speculum was cast, and of conversing with the workmen in the grounds, some of whom had assisted in the construction, and who were amusingly proud of "her," as they designated the telescope. It was a delight to him also to see the crowds of happy and wondering faces that gather round this scientific marvel —the whole grounds being thrown open both to townspeople and strangers every day after two o'clock; they are also permitted to examine the works, and to walk through the telescope. A few contemporary recollections of our surroundings as well as of our first night of observation may be interesting to unscientific readers :—

" Parsonstown, or Birr, is a town in King's county, so comfortable and so pleasantly situated, with its rows of old trees, its pretty church, its flourishing shops and villas, that I almost supposed myself to be in a Scotch or English town. Not a hundred yards from it, and divided from one of its streets only by an immense castellated and ivyed wall, stands Birr Castle, the ancestral abode of William Parsons, Earl of Rosse ; it dates from 1620, but in 1832 a fire destroyed the centre part of the edifice, which has been replaced in excellent keeping with the old building. The country around is flat, with the exception of a curious conical hill, which terminates one of the park vistas, and is the famous Tipperary hill, well known in fairy lore as Knocksheogowna, ' the hill of the fairy calf.' Close to the castle windows flows the Blackwater, which joins the Brusna, about an Irish mile from the house, forming at its junction the boundary between King's county and Tipperary. These two rivers exhibit the phenomenon of the Rhone and Arve—the muddy Brusna flowing on for some distance beside the clear tide of the Blackwater, without mingling their currents. In front of the castle extends a wide and beautiful demesne, in passing through which the eye is struck by two remarkable objects ; a white building consisting of two parallel walls, between which is observed something huge, dark, and tower-like ; while a few yards further on stood a large structure of black legs and arms, looking very like a spider of the mammoth and mastodon period. The first is the six-feet mirror telescope, calmly reposing in its unconscious fame; the second is the three-feet mirror telescope, which, though eclipsed by its stately rival, yet possesses one material advantage, as we shall afterwards see.

" As Irish skies are proverbially not the clearest in the world, our hopes were very faint of seeing what we came to see; and a tradition of Sir James South having been detained for six weeks without one clear night, grated unpleasantly on the ear. People who repose quietly in their arm-chairs after dinner, and go to bed at reasonable hours, have small idea of the excitement prevailing in an astronomer's household as to the state of the weather. How anxiously we all watched the dark banks of cloud, and the tremulous fleecy vapour upon the blue sky, and the rich crimson, gold, and green of the sun-setting! At last,

'When the sun fell, and all the land was dark,'

the welcome summons reached us from the observatory. Truly, to approach the giant telescope for the first time by night is a scene never to be forgotten; but faintly indicated by the dim star-light and partially veiled moon, the immense structure loomed almost awfully in the obscurity. On one side, the open door and windows of the small observatory, which contains two transit instruments, gave out a gush of light and warmth; on the other side, the black and unenclosed scaffolding of the three-feet telescope stood out against the dark blue heavens. Although the skies were looking down upon a turbulent land, the only sounds that broke the silence of midnight were the whisperings of the lake, the river, and the trees.

" Between two large piers or walls is suspended the
great telescope.  Upon reaching the top by a slender
staircase, we were introduced into a small but apparently steady gallery.  The telescope was pointed to
the heavens, about twenty feet from where we stood ;
beneath us was a depth of sixty feet, and no apparent
way of bridging the chasm.  At last one of the assist-
ants, stationed a few steps below, turned a small wind-
lass, and lo ! we began to move gently through the air
till we arrived at the mountain-like side of the telescope, about four or five feet from the mouth.  With
no supports from beneath, it appeared as if we were
poised in mid-air; strong wooden beams, however,
secured by iron slides, supported the gallery from the
wall which we had just left.  So imperceptible is the
motion, that one night a gentleman, unconscious of
having left the solid landing-place, opened the gallery
door, and walked sixteen or twenty feet with no other
footing than a narrow unrailed beam, but, almost mira-
culously, he reached the other side in safety, although
he fainted the next morning on being taken to see the
escape he had made.

" The six-feet concave mirror or speculum is made of
tin mixed with copper, and polished to an exceeding
brightness.  In looking into the mouth of the telescope
by what is called the front view, we see the inverted
image formed by reflection from this mirror in tremulous and dazzling radiance, but it is not thus that ordinary observations are made ; a second mirror of small
size is placed at an angle of 45°, so as to reflect the
image to the side of the instrument, where it is viewed
through eye-pieces of different magnitudes.  We took
our places at this point by aid of the aforesaid ' aërial

machine.' The evening, though lovely to unastronomi-
cal eyes, was not altogether favourable for observations;
however, we saw

> ' The galaxy, that milky way,
> Which nightly as a circling zone thou seest
> Powdered with stars.'

" It is impossible to describe the distinctness, and the
nearness, and the individuality of the ' starry powder ; '
in the middle of it was a double star [always a peculiar
object of interest and study to my father]—' twin suns,'
as they have been called, ' moving in their mysteriously
united beauty and brightness.' Of course the great
object of ambition was to see the nebulæ—the resolution of which by the giant tube destroyed that plausible
theory, which when carried to its greatest extent made
such dangerous aggressions against the divine creative
acts, originating all worlds from a slowly progressing
vapour and fire mist. The atmosphere would only allow
us to see the Dumb Bell Nebula, so called from a supposed resemblance in its form to that instrument ; it is
only a partially resolved one, however, even by the large
instrument—the portion of white vapour which is still
observable, though

> ' Sown with stars, thick as a field,'

will require some higher process to be yet developed.

" We could not see the moon through the six-feet
telescope, as she was not within the meridional range,
which in this instrument is limited by the two walls ;
the disadvantage of which is counterbalanced by the
speed and steadiness with which it can be lowered or
elevated. We accordingly repaired to the three-feet
telescope, which can be pointed to all quarters of the
heavens, and to which the following high comparative

praise has been given :—'To look through Herschel's four-feet mirror, compared with Lord Rosse's three-feet mirror, is like a short-sighted person looking at the stars without his spectacles.' [The ladder of ascent was a very precarious one, and my father missed his footing in the imperfect light, narrowly escaping serious injury. It was the second accident of the kind which he had met with—the first having occurred, I believe, about forty years before, when examining a large telescope belonging to Mr. Ramage of Aberdeen.] Once ascended in safety, we gazed with wonder upon the lunar valleys, mountains, and caverns, so near and so distinct, that there seemed no obstacle to taking a quiet walk amidst their lights and shadows, their deep ravines, their volcanic cones and cavities. The silence and the immobility of that bright world was almost oppressive ; one gazed and listened, expecting to see and hear life, but no life was there. It is believed that if there were large buildings like a church, or a mill, or railway works, they could be clearly discerned ; but there has been no change or furrow since human eyes were permitted to draw nigh its calm surface. Were there inhabitants, they must be independent of air and water, and must be scorched in light and heat one-half of each month, and frozen in cold and darkness the other. The whole scene, even in the intervals of active vision, was one suggestive of much solemn thought. Every now and then a meteor flashed excitedly amidst the calm stars and planets, and even as swift and short seemed the career of man in comparison with the ages of the past and future, chronicled before us in the heavens. 'Hither shalt thou come and no further" seemed legibly written before the genius of man ; while all that he yearns to know, and cannot

know, is known, it may be, to the babe that has gone
but yesternight to glory."

Astronomy was not the only subject which interested
my father at Parsonstown.    Its Presbyterian pastor,
the late Rev. Dr. Carlisle,[1] had given up a good living
and a highly respectable congregation in Dublin for
the purpose of becoming a missionary among the Roman
Catholics of Birr.    This self-denial being the spirit
of the primitive church, was honoured of God, as
self-denial ever is, and he was at this time carrying
on most successful mission-work in Parsonstown.    It
was, alas ! a " season of landlord shooting," and the
south and mid counties of Ireland were in a most
turbulent state.    The insecurity of life from day to day
even of kind and faithful landlords, and the many localities in these disturbed counties which were signalized
as scenes of painful deeds, could not fail to impress the
minds of strangers.    Yet this man walked in peace and
in favour through the streets where a few years before he
had been stoned and persecuted.    Truly the missionary's
feet, though not shod in silver, were enabled to walk
in ease over the difficult places like the feet of the hind
upon the rocks.    My father attended Dr. Carlisle's
preaching with great interest, and was taken by him
to see his schools and missions, and also went with him
through several convents in the town.

We next proceeded to visit Mr. King of Ballylin,—
nearly related to Lord Rosse,—in King's county, where

[1] The work at Birr has languished considerably since the death of this
self-denying missionary, but the school is still carried on, principally under
the superintendence of Dr. Wallace, a medical man there, son of the Rev.
Dr. John Wallace, previous to the Disruption, minister of Abbey St.
Bathans.    He takes photographs of Birr Castle, the Telescope, etc., and
by their sale helps the funds of the school.

we witnessed a beautiful specimen of the life and residence of a genuine " old Irish country gentleman," the
acquaintance with whose family we had made at Birr
Castle. One of the daughters of this hospitable home,
so green and lovely among its stately beech trees, was
for years a correspondent and a kind assistant to my
father, having great, though unpretending, scientific acquirements, and, moreover, an exquisite gift of drawing,
which she turned to good account in illustrating his
scientific papers as well as her own charming books.[1]

These kind friends took us, among other sights of the
neighbourhood, to Clonmacnoise, a collection of ruins
on the banks of the Shannon. This scene deeply impressed itself upon my father's mind, and was often
referred to by him. It included the remains of seven
churches, two round towers (one of them, by a curious
coincidence, being the exact size of Lord Rosse's telescope), two old ruined crosses and a " Holy Well," and
presented plenty of materials for sketching, legendary
stories, and antiquarianism ; but a sadder and a stranger
sight was there. Clonmacnoise was a " pilgrim sta-

[1] Since writing the above, a melancholy event has caused the deepest
sorrow to a large circle of attached relatives and friends, besides desolating one of the happiest of homes. By a most sad accident in the street
of Parsonstown, the above-mentioned lady (the Honourable Mrs. Ward)
met with an instantaneous death, which to her was, however, truly an
entrance into life. She united in a singular degree great and varied
talent with extreme simplicity, sweetness, and humility, and in every relation of life peculiarly " adorned" the Christianity which was her hope and
joy. Of her it was said that she was "the gilder of every dark cloud" to
those around her. She went the day before her death to visit Lord Rosse's
vault, in order to copy the inscription on his tomb, and, by a singular
coincidence, the first name that caught her eye on entering it was her own
maiden one, " Mary King," on the plate of one of the tombs. Mrs.
Ward was the author of *Telescope Teachings*, a new edition of which had
just been called for, also of *Microscope Teachings*, and of many articles
in the *Intellectual Observer*.

tion," and there groups of poor women, some literally
clad in tatters, others in the picturesque red petti-
coat which marks the Galway peasant, were painfully
stumbling over the rough ground on their knees, tying
bits of ribbon to crosses, and vowing and performing
vows to the Virgin and saints, in our own English
language and within our own British realm. Our next
move was to Dublin, by the way of Maynooth, where
my father stopped to see a scientific *confrère*, one of the
professors. We were kindly and hospitably entertained
at luncheon, and shown through the noble pile of
buildings with all its paraphernalia of an enlightened
education for Irish priests, carried on at the expense of
Great Britain. The recollections, however, of the pilgrim
station of the day before, fully sanctioned, as we had
ascertained, by the parish Roman Catholic priest, could
not fail to intrude upon Scottish minds. From Dublin,
with its striking contrast of wealth and misery, of
squalid hovels and stately squares, we went again northward to attend the twenty-second meeting of the
British Association at Belfast, where we entered on yet
another phase of life in Ireland,—that of its princely
merchants,—being entertained with Irish hospitality at
Hopefield, the beautiful country villa of the late J.
Sinclair, Esq., then thrown open to strangers attending
the Association. The illness which had hovered round
Sir David ever since he entered Ireland here assumed
a very serious aspect, and he was scarcely able to attend
any of the meetings. The sea voyage and Scotch air
speedily re-established him, however, in nearly his
usual health.

The Emperor of France bestowed on him in this year
the decoration of the Legion of Honour. He was also

chosen President of the Working Men's Educational Union,—a fitting position for one who has been called "the people's philosopher;" so much had he done to utilize science and to make it popular.

When in London in the spring of 1853, we have the following accounts of a few of the noted persons he was meeting day by day :—

"LONDON, *May* 1853.

"I had the good fortune to meet Mrs. Beecher Stowe at Mr. Rogers' on Tuesday. There was only a small party there,—Lord and Lady Hatherton, Lord Glenelg, and a few more, so that we saw her to advantage. After her visit to Mr. Rogers, she was engaged to visit the Bridgewater Gallery of Pictures, and I was invited to accompany the party. In the absence of Lord Ellesmere, who has gone with his family to America, as a Commissioner to the Great Exhibition of New York, his brother, the Duke of Sutherland, accompanied by the Duchess, did the honours of the gallery to Mrs. B. Stowe. I thus saw a good deal of the lady. She looks young, with a short figure, but a highly intellectual expression of face, particularly in the eyes and forehead. She was quite at her ease, and seemed no way spoiled by the notice she received. The greatest compliment she has yet received is from this week's *Punch*, where the principal picture represents her and Diogenes, with the inscription, 'Diogenes has long sought in vain for an honest man, but having now found an honest woman, has extinguished his lamp in honour of Mrs. Beecher Stowe.'"

"LONDON, *May* 28, 1853.

"On the 26th I breakfasted with Dr. Milman, where we had a brilliant intellectual party,—Macaulay, Hallam,

Dr. Hawtry, head-master of Eton, and the ex-President of the United States, Van Buren, who was astonished at the conversation, which Macaulay generally guides, —the subject happened to turn on some subjects connected with Mental Philosophy, on which I was at home, and I got my full share in the discussion, in opposition to Macaulay's speculations.

" ATHENÆUM, *June* 25, 1853.

" I went next day, and took Lady and Miss Becher with me, to the Reformatory School at Westminster, where there was a meeting of the trustees, on the occasion of seventeen of the reformed persons going out to America. Lord Shaftesbury and several gentlemen made interesting speeches. This school, called the place of reformation, is open to thieves of all kinds (males, I believe, only) who choose to come, and they may go when they please. They receive only a certain number, perhaps thirty or forty, and there have been 3000 applications."

# CHAPTER XV.

## NOTES OF LIFE IN 1854-55.

A SPECK to *us !* a world to those
  Who bask within its sphere ;
Whose sun-bright sky with ardour glows,
  That would consume us here !

For what are matter's noblest forms
  If mind be wanting there?
A chaos REASON'S ray ne'er warms,
  Were dark midst brightest glare.

And He who gave us *her* blest beam,
  In this dim world afar,
Would ne'er deny her radiant gleam
  To that bright glowing star !

Would ne'er amid such dazzling light
  A world of glory place,
And leave it lost in mental night
  Its aimless course to trace !—ETA.

THE year 1854, although one of most engrossing work, commenced with prostrating illness, in the form of a long low influenza. Before he had recovered from it, he received a request from Professor Fraser, then the editor of the *North British*, to review a certain anonymous volume entitled *Of the Plurality of Worlds*, an essay which was exciting much attention. All illness was forgotten, and he was soon in one of his entirely absorbed states upon the subject. The severity of the review has been ascribed to a certain amount of personal feeling which existed between the author and the, critic, owing to some passages of arms at "the tournament of science," at various times and places. This may afterwards have added a shade of severity to his satire, but he had fully made up his mind as to the merits of the argument and the volume, before he knew who had written it. I make the following extracts from two contemporary letters :—

*"Feb.* 4, 1854.

" Tell me anything you know or can collect about ' the Plurality of Worlds ;' he [Sir David] has been particularly requested to review it, and is going over it just now, groaning at every line ; he says it is ' quite disgusting,' and displays great ignorance ; he wants very much to know who is the author. He has not finished the perusal, and may be in better humour with it before he closes. You will make allowance for his strength of expression ! I must say, however, that the passages he has read to me are rather weak. We are also reading Fontenelle's little book on the same subject, or rather the same name, more honestly used, and it is very amusing, quaint, sparkling, and vivacious."

*"Feb.* 10.—Many thanks about the ' Plurality.' We have since heard from authority that it is by Whewell !!! which is surprising, as his views in his Bridgewater Treatise seem rather different."

To those behind the scenes, it was abundantly evident that the personal depth of feeling which he displayed upon this occasion arose from his characteristic liveliness of participation in any subject which deeply touched him. The eye and the mind, accustomed from early childhood to gaze out at the beautiful worlds rolling above, felt personally injured by the dreariness and narrowness of the views which he combated. Sir Isaac Newton had written long before,—" For in God's house (which is the universe) are many mansions, and He governs them by agents which can pass through the heavens from one mansion to another. For if all places to which we have access are filled with living creatures, why should all these immense spaces of the heavens above the clouds be incapable of inhabitants?" And his disciple had clung to the same belief since the days of his youth.

I shall never forget the delight and satisfaction with which, in the course of his own private study of the Bible, he came upon this verse in Isaiah,—"For thus saith the Lord that created the heavens, God himself that formed the earth, and made it; he hath established it, he created it not in vain, he formed it to be inhabited;"[1] which his mind at once seized with ardour, as a logical demonstration that the others, if uninhabited, would have been "created in vain." In fact, as he studied this subject, the habitation of " More worlds than one " by intelligent creatures seemed impressed on his mind as if written by a sunbeam; and although in the midst of his greatest work, the extended biography of Sir Isaac Newton, he felt impelled to lay it aside for a time, and bent all his energies in preparing a volume with the above happy title. Rapidly his occupied mind poured itself upon paper, and the work was soon in proof. Just at this time we proceeded to Clifton, on the occasion of the marriage of a young friend and connexion to whom he was much attached.[2] We visited, *en route*, his friend and correspondent, Mr. Chance, of the great glass-works at Birmingham, where

[1] Isaiah xlv. 18.
[2] Miss Charlotte Heriot Maitland, the wife of Frederick King, Esq. of Fryern.

he was greatly interested in seeing the English and French methods of blowing glass, the latter with its inventor, M. Bontemps, as cicerone. It was at No. 9 Princes Buildings, Clifton, where we were with his old friends and neighbours of Coates Crescent days, Mr. and Mrs. George Forbes, that the proof sheets of *More Worlds than One* were corrected. There, in an invalid room, he read day by day his proofs to me and a favourite young friend of his,[1] who sends me the following recollections of a time, when the humility of the great mind was very touchingly manifested :—

"I may now hope to put down a few long-cherished reminiscences of that dear father of yours, whom I loved so well, and in whose society we all so delighted. I am only sorry that my bad memory does not retain many more vivid impressions and traditional stories, having enjoyed so many opportunities of intercourse with him, and having heard so many characteristic traits of him from my own dear father. One of the episodes in our intercourse which I remember most clearly, occurred that spring we all spent at Clifton. Sir David was then writing his *More Worlds than One*, and he asked you and me to help him in correcting the press. In the course of this most interesting work we came across several expressions we thought much too severe, and we summoned courage to point these out to the learned author—who at once altered them in the meekest way to our entire satisfaction ; there was, however, one whole sentence which we much objected to— your father said he had looked at it ; he did not see anything so very objectionable about it ; he did not think

[1] Miss Forbes, now the wife of the Rev. Canon Harford Battersby, St. John's Parsonage, Keswick.

he could put it differently ; it was printed, so it must just stand, otherwise there would be a blank in the page, which would never do. We still ventured to persist that we did not think the passage could remain as it was, upon which he said, half provoked and half amused at our audacity and pertinacity, that we were welcome to strike it out if we could write a paragraph to fill the space ; this we accordingly did, and inserted something which was at all events an improvement in point of amenity ! Another thing I remember in connexion with this subject, is my mother gently remonstrating with Sir David in regard to the somewhat unmeasured terms in which he spoke of the author in his review in the *North British* of the *Plurality of Worlds*. She said such expressions were calculated *to hurt his feelings*. ' Hurt *his* feelings ! ' broke in Sir David, ' why, it is he that has hurt *my* feelings ! '

" All who knew him will, I am sure, unite in testifying to his readiness to explain, it might be, the simplest principles of a science to some insignificant person, and the wonderful enjoyment he seemed to find in so doing, quite as much indeed as in talking of some of his latest discoveries to the most learned,—if only his listener were thoroughly interested, and anxious to learn. In illustration of this, I may mention that my dear mother says she has two sheets he wrote out for her many years ago, explaining some scientific point she had had some difficulty in understanding.

" It was delightful too to observe the fresh admiration, delight, and wonder which he himself felt, each time he spoke of or exhibited some of the exquisite and infinitely small works of creation as revealed by the microscope, or the infinitely large and splendid

heavenly bodies as presented to us by the telescope.
And now how blessed it is to think that, with all his
acquired learning, and his many marvellous discoveries,
above and beyond all these—which he prized so much,
and which he had laboured perseveringly throughout
his whole life to attain—he rejoiced most of all in the
' *knowledge* of *Christ* and in the power of His resurrection.' "

*More Worlds than One* became a very popular volume,
and produced a flood of correspondence,—many quota-
tions, original poems, and notices of old books being
sent to him.   He wrote on one occasion :—" Mr.
Monckton Milnes gave me yesterday the following
poetical translation of a fine sentiment of Immanuel
Kant's for another edition of *More Worlds :*—

> ' *Two* things I ever tremble when I scan—
> The star-lit heavens, the sense of Right in man.' "

Another sends, for " next edition," the following quotation from Samuel Rogers :—

> " Now the day is spent,
> And stars are kindling in the firmament ;
> To us how silent—though, like ours, perchance,
> Busy, and full of life and circumstance."

Another, the following from Sir E. B. Lytton-Bulwer :—

> " Can every leaf a teeming world contain,
> Can every globule gird a countless race ;
> Yet one death-slumber in its dreamless reign
> Clasp all the illumed magnificence of space ?
> LIFE crowd a grain, from air's vast realms effaced ?
> The leaf, a world—the firmament, a waste?"

And he received the following letter :—

" REFORM CLUB, 17*th May* 1860.

" DEAR SIR,—I have met with so very amusing a
proof of the *non-plurality of worlds* in a fusty old book,

that I cannot resist the temptation of sending it to you, as it is probably unknown. Two French clergymen were arguing the question on theological grounds, and one of them took refuge in the parable of the lepers (Luke xvii. 17), quoting the Latin version, 'Erant decem MUNDI;' to which the other answered by continuing the quotation, '*ubi sunt reliqui novem.*' If Sir Thomas Browne had written on this subject I am sure he would have trotted out this argument with great energy ; at least in the humble opinion of, truly yours,

" C. DE LA PRYME."

It is right to state that these two knights of science, true and brave, Brewster and Whewell, both now passed beyond earthly conflict, were thoroughly reconciled to each other some years after this severe test of literary friendship.

After some time spent in London, we went to Leamington, to visit Dr. and Mrs. Burbidge, where my father enjoyed much congenial society. Among his many interests there, were Mr. Craig, the constructor of the large achromatic telescope, and Mr. Buckle, the photographer ; it was a great pleasure to him also to see his old friend Dr. Jephson, who, although in blindness and feebleness, and thus enforced leisure, still was in possession of his old vivid life.

Thence we went to Manchester, paying a most interesting visit to Mr. Fairbairn, which was shared by the late Mr. Hopkins, of Cambridge, the celebrated private tutor in mathematics, whose society always afforded my father peculiar pleasure. No sight, no kind of information, ever came amiss to the latter, who was, as usual, deeply engrossed in "examining" the Manchester factories, the locomotives, and engineering improvements of his host, and the steam-hammer and numberless curious experiments and inventions of another Scotch friend, Mr. James Nasmyth. One day's expedition he counted as "a white day" of his life. It was a visit to Saltaire in company with Mr. Fairbairn, his accomplished daughter Mrs. Bateman, and Mrs. Gaskell, the popular authoress, now, alas! no more. He was deeply interested in the alpaca factory, in the flourishing flock of alpacas, in the model town of 1000 workers and their families, in which was church and school, and not one public-house, and, most of all, in Mr. (now Sir) Titus Salt, the creator and proprietor of all this well-regulated power and wealth, and in his beautiful and refined home, "The Crow's Nest."

In August of the same year we went for the first time to Aberdeenshire, afterwards to be a place of many tender and happy memories of him. We first visited Keith-hall, the seat of the Earl of Kintore, whence he was taken to see many of the fine old castles for which Aberdeenshire is celebrated,—Fyvie, Castle Forbes, Castle Fraser, etc., which he enjoyed much, and also made the acquaintance of Lord Aberdeen, the statesman, at Haddo House. The next visit was to Banchory House, where his previous acquaintance with Alexander Thomson of Banchory ripened into a warm and sympathetic friendship.

In the spring of this year the letters forming his London journal turn very much upon spirit-rapping and table-turning. As these letters give his vivid impressions as thrown off at the time, I give them together, although of different dates. Regarding a kindred subject, the following quotation well illustrates

the lively interest, mingled with scientific caution, with
which he treated these topics.   He went down to
Brighton to see the curious experiments connected with
the magnetoscope, and I find a rough draft of his opinion of these experiments with this note :—" This was
drawn up at the desire of Mrs. Lee of Hartwell, who
wished my opinion of the Magnetoscope, and of Mr.
Rutter, who was, I believe, her nephew."

   " On the 6th of August 1851, I was requested by Lady
Byron and Lady Lovelace to accompany them to Mr.
Rutter's, at Brighton, to see his experiments with a pear-
shaped ball of wax suspended by a fibre of silk.   The
experiments were very successful, and from Mr. Rutter's
character and talents, we were sure that they were
honestly performed.   The general result was that the
waxen ball revolved from left to right when held by a
male hand, but stopped and revolved in the opposite
direction when any article belonging to a female was
laid on the hand or arm of the male.   Various changes
on the motion of the ball were produced by animal
substances, by light, galvanism, and electricity.   It
would be desirable to have these experiments carefully
repeated when *the operator is blindfolded*, for there is
reason to believe that when there is a previous knowledge of the effect to be produced, a desire that it should
be produced may involuntarily influence the mind of
the operator, and that this desire as involuntarily may
influence his hand to give such an impulse to the
suspended line, as may be required to produce the
expected result.   Mr. Rutter, indeed, showed us an
experiment in which his hand was separated from the
suspending thread by a fixed arm of glass and metal,
supported by a stand ; but the effect was too evanescent

to entitle us to regard the result as a scientific fact upon which implicit reliance could be placed.

DAVID BREWSTER."

" HARTWELL HOUSE, *Oct.* 21, 1851."

" LONDON, *May* 1851.

" I HAVE been at two mesmeric *séances*, one with Dr. Macdonald and the Duke of Argyll, at a Mrs. Holmes', who failed utterly in her clairvoyante pretensions. A Count Possenti mesmerized her. The other was at Dr. Ashburner's, where I saw things that confounded me."

" LONDON, *April* 25, 1851.

" WE had really a delightful breakfast party at the Chevalier Bunsen's, one of the most learned men of the age, and so frank and kind. The great subject of talk here is spirit-rapping, and the moving of tables; when the party sitting round the table place both their hands upon it, the thumb of the left hand touching the thumb of the right, and the little fingers touching the little fingers of the hands of the persons on each side of you, it is then said to shake and tremble, and often to be moved along the floor. Just as we were discussing the subject, Mr. Bunsen received a letter from the King of Prussia, saying that the experiment was made at the Palace by the royal party, who were alone, and no conjuror present. Three of the young ladies had each letters from Berlin, mentioning these experiments, which sometimes fail. One letter stated that it succeeded three times out of seven. Another account described the size of the table, the wood of which it was made, and all the particulars of the experiments, with the greatest minuteness. Of course it is nonsense, and there must be some trick in it. . . .

" . . . Mr. Monckton Milnes asked us to breakfast with him to-morrow, to meet Mr. Galla, the African traveller, who assured him that Mrs. Hayden told him the names of *persons* and *places* in Africa which nobody but himself knew. The world is obviously going mad. An American whom I met at Rogers' the other day told me that hundreds had been sent to lunatic asylums in consequence of the communications made to them by the spirit-rappers. . . . The spirit-rapping is exciting great interest in London, but very few believe in it, and there are many facts which tend to prove that it is done by some machinery or apparatus by which the hands and feet of the medium may produce the observed phenomena.[1] . . .

" In the table-turning the table moves round, and you

[1] Singularly corroborative of this opinion is the following letter, which appears in the *Standard* of October 7th, 1869 :—

" SIR,—For many years I have had a large sale for spirit-rapping magnets and batteries expressly made for concealment under the floor, in cupboards, under tables, and even for the interior of the centre support of large round tables and boxes. I have supplied to the same parties quantities of prepared wire to be placed under the carpets and oilcloth, or under the wainscot and gilt beading around ceilings and rooms ; in fact, for every conceivable place. All these obviously were used for spirit-rapping, and the connexion to each rapper and battery was to be made by means of a small button, like those used for telegraphic bell-ringing purposes, or by means of a brass-headed or other nail under the carpet, at particular patterns known to the spiritualist. I have supplied all these in considerable numbers, but of late years the demand has ceased, owing, I trust, to the march of intellect which has exposed the imposition. These rappers, when carefully placed, are calculated to mislead the most wary. Then there are spirit-rapping magnets and batteries constructed expressly for the pocket, and, of course, these will rap at any part of the room. I have also made drums and bells which will beat and ring at command ; but these two latter are not so frequently used as the magnets are, because they are too easily detected. . . .—Your obedient servant,

" W. G. FAULKNER,
Philosophical Instrument Maker."

" 40 Endell Street, W.C., *Oct.* 6."

are obliged to follow it. It often runs away !! and it is now found that it *obeys* the commands of one of the movers, and tells secrets to him when asked. The Chevalier Neukomm, who inquired in the kindest manner after you, told me that the table when questioned told his age. One of the party desires it to do this by lifting up one leg, and rapping just one rap for every ten years, then quick raps for every unit. He pressed me to go and see this done at a private party, where he was to dine with a scientific medical friend. I went, and saw and heard the table do all these wonders. It told my age, but blundered a little. Now all this was done by an involuntary action of the fingers of the party. My hands were on the table, and I could perceive no trick on the part of the others there. . . .

"I believe the truth to be this. Electricity and magnetism have nothing to do with it. Neither the one nor the other can pass from the body unless by a strong muscular effort, and in that case it requires the most delicate galvanoscope to make it visible. But even if there was an abundance of electricity in the body, it could not enter the table, which is a non-conductor, and, even if the table were a conductor, its effect would not be to turn it. When a number of hands are so placed, either in contact with one another or not, there is necessarily a tremulous motion from the circulation of the blood, and the fatigue of remaining in one position ; and when this motion is communicated to the table, and the party wish it to move, as they are directed to do, from right to left, they involuntarily help it forward, while following its first motion. In this way it succeeds with most operators. I have no doubt that there are *thou-*

*sands* of tables turning every night in London, so general is the excitement on the subject."

"LONDON, *June* 1855.

"Last of all I went with Lord Brougham to a *séance* of the new spirit-rapper, Mr. Home, a lad of twenty, the son of a brother of the late Earl of Home. He went to America at the age of seven, and though a naturalized American, is actually a Scotchman. Mr. Home lives in Coxe's Hotel, in Jermyn Street; and Mr. Coxe, who knows Lord Brougham, wished him to have a *séance*, and his Lordship invited me to accompany him, in order to assist in finding out the trick. We four sat down at a moderately-sized table, the structure of which we were invited to examine. In a short time the table shuddered, and a tremulous motion ran up all our arms; at our bidding these motions ceased and returned.

"The most unaccountable rappings were produced in various parts of the table, and the table actually rose from the ground when no hand was upon it. A larger table was produced, and exhibited similar movements.

"An accordion was held in Lord Brougham's hand and gave out a single note, but the experiment was a failure; it would not play either in his hand or mine.

"A small hand-bell was then laid down with its mouth on the carpet, and after lying for some time it actually rang when nothing could have touched it. The bell was then placed on the other side, still upon the carpet, and it came over to me and placed itself in my hand. It did the same to Lord Brougham.

"These were the principal experiments; we could give no explanation of them, and could not conjecture how they could be produced by any kind of mechanism.

R

Hands are sometimes seen and felt, the hand often grasps another, and melts away, as it were, under the grasp.

" The object of asking Lord Brougham and me seems to have been to get our favourable opinion of the exhibition, but though neither of us can explain what we saw, we do not believe that it was the work of idle spirits."

# CHAPTER XVI.

## NOTES OF LIFE FROM 1855 TO 1860.

" Here Newton dawned, here lofty wisdom woke,
And to a wondering world divinely spoke."

" A chief of men who through a cloud,
Not of strife only, but detractions rude,
Guided by faith and matchless fortitude,
To peace and truth thy glorious way hast ploughed,
And on the neck of crowned science proud,
Hast viewed God's trophies, and this work pursued."

IN the summer of 1855 my father published the *Memoirs of the Life, Writings, and Discoveries of Sir Isaac Newton,* which was dedicated to Prince Albert, not only as Chancellor of the University of Cambridge, " the birthplace of Newton's genius, and the scene of his intellectual achievements," but as a Prince of such noble and patriotic views, who had given such an impulse to the arts and sciences of England, that if fully seconded the result must have been " a complete national encouragement of science and consolidation of our scientific institutions."

Brewster's admiration of Newton dated from an early age. There is a tombstone on the south wall of Grey-friars Church, Edinburgh, erected in memory of Colin Maclaurin, Professor of Mathematics in Marischal College, Aberdeen, on which is recorded in two words Newton's recommendation of him. My father thus writes in his Life of Sir Isaac :—" When a youth at College, I have often gazed upon this simple monument, and pondered over the words, to be envied by every aspirant to scientific fame, 'NEWTONE SUADENTE.'" In 1814 he visited the Manor of Woolsthorpe, the birth-place and paternal estate of Sir Isaac,—saw the celebrated apple-tree, which was said by one of its falling apples to have suggested to Newton the laws of gravity,—and brought away a portion of one of its roots. There was much similarity between the genius, the characteristic individuality, and the career of both. Had they been contemporaries doubtless there would have been mutual warm personal sympathies. As it was, there was some-thing approaching to the known and the personal in the affectionate admiration which Brewster ever cherished for Newton. Four years after he wrote the shorter Life, several statements deeply affecting the character of the great author of the *Principia* had been given to the world in a work of great interest, *The Life of Flamsteed*, by Baily, printed and circulated at the expense of the Board of Admiralty,—which were very injurious to the fame of the philosopher.

> " Gnats are unnoticed wheresoe'er they fly,
> But eagles gazed upon by every eye ; "

so that Newton had had his full share during life, of controversies, accusations, and irritating scientific feuds, but had come out of them all as the acknowledged chief of English science. That this great man should be attacked more than a hundred years after his death, was to Brewster's mind a personal grief and an English scandal. He therefore for twenty years made it one of his objects in life to search out every proof and evidence by which he could defend Newton from the charges against his sanity, his probity, and his justice, which were circulated when the hand and the tongue of the accused and his contemporaries were safely mouldering in the grave. In 1837 Sir David "applied to the Hon. Newton Fellowes, one of the trustees of the Earl of Portsmouth, for permission to inspect the MSS. and correspondence of Sir Isaac Newton, which through his grand-niece, Miss Conduitt, afterwards Lady Lymington, had come into the possession of that family." Mr. Fellowes kindly granted this request, and aided by his son, the late Mr. Henry Arthur Fellowes, my father had the fullest opportunities at Hurtsbourne Park for collecting entirely fresh material for his great work. Mr. Baily had mentioned "the valuable collection of Newton's MSS. belonging to the Earl of Portsmouth," but "stated that he found nothing in it to throw any light on the special object of his inquiries," and in his book published only eleven of Flamsteed's letters to Newton. My father, however, found forty in addition to those, which he considered of great importance in judging of the difficult and delicate controversy between these two distinguished individuals, and in which he believed that Newton had been grievously traduced. There was another point on which Brewster's heart was set on vindicating his unseen friend and master. Of undoubted orthodoxy himself, he could not endure that Newton should be considered a Socinian or Unitarian, and in his first work had not scrupled to declare him to be a Trinitarian. Ampler information, however, came before him, and he considered it a duty to investigate the matter more fully. I remember his delight when he came upon passages clearly showing, as he thought, that Sir Isaac was neither a Socinian nor an ordinary Arian, and his mortification was deep when he could not find any word or note in Newton's writings that could prove him to be beyond what is, I believe, called an "advanced Arian." He held that Jesus " was the Son of God, as well by his resurrection from the dead as by his supernatural birth of the Virgin," although there is never any recognition of His equality with the Father. His biographer always believed, however, that the proofs were negative, and seemed to cling to what he considered a fact, that there was no distinct declaration of Newton's rejection of the doctrine of the Trinity. He lived in communion with the Church of England, and studied and loved the Word of God. He was described by one bishop as "knowing more of the Scriptures than them all," and by another as having "the whitest soul he ever knew;" and while we must grieve that he had not grasped the whole of Scriptural truth for himself, yet it is delightful to find this great man of science warmly interested in the welfare of immortal souls. He was often consulted about the spiritual state of his friends, and prayed for them. One eminent mathematician " thanked God that his soul was extremely quiet, in which Newton had the chief share;" and Dr. Morland, a Fellow of the Royal Society, wrote, " I have done, and will do, my best while I live to follow your advice to repent and believe." The whole of this book, though. it involved severe labour, was most congenial work to Brewster. He especially enjoyed the correction of its sheets, feeling then that the heaviest part was accomplished, and he corrected every line with peculiar and loving care.

Working in the same demesne, and familiar with the marvellous discoveries of Newton, it seemed as if it were with reluctance that Brewster admitted of errors either of omission or commission in those great labours. Although obliged to admit, by stern scientific facts, the superiority of the undulatory theory of light, yet it seemed as if his mind lingered reluctantly over the beautiful Newtonian doctrine of emission. He evidently grieved that the elder philosopher had missed the great discovery made by Wollaston and Fraunhofer of the black lines in the Solar Spectrum, upon which he himself so improved, by increasing the 600 previously discovered to upwards of 2000 by patient observation and the use of excellent instruments. It is beautiful to observe how these and other masters of science, each in his day, add a quota of riches to the treasure-house of discovered knowledge. Newton's theory of the sevenfold colours of the spectrum, which he likened to the seven notes of the musical chord, was of value in its time, although reduced by Brewster to the primary colours, blue, red, and yellow, which again did good service, although that again may be considered as disproved by the more recent researches of Clerk-Maxwell and Helmholtz. Brewster's mind clung, however, to his own trinity of colour, which he never gave up. He carried it into his study of pictures, criticising severely some of the ancient masters and their disciples for their utter neglect of what he considered the scientific harmony of colouring. A pretty experiment showing the tremulous movement of a red spot upon a green ground, or blue upon orange, was a great favourite; he exhibited it as shown by a mat of worsted work at a meeting of the British Association, explaining the phenomenon by his favourite theory of colours. Blue and red, yellow and green, without being softened by a due proportion of their complementary hues, seemed to give positive physical pain to his sensitive vision, whether in dress, paintings, or furniture. Upon one occasion he resolved to bring home a present which should be at the same time a scientific lesson. The result was a dress in which red had its sufficient complement of green, and blue its proper companion, orange. Unscientific eyes were compelled to grieve that it appeared also to possess the uncommon quality of never wearing out !

In the summer of this year, Sir David Brewster was chosen a juror of the Paris Exhibition for the department of Optical Instruments, and he spent much time and energy in the fulfilment of his congenial duties. He wrote thus :—

"Paris, 34 Avenue d'Antin, Champs Elysées, *Wednesday, July* 11*th*, 1855.

" I left London on Monday at 1.30, and got on board the packet-boat at Folkstone at 4.30, the wind, as I thought, blowing a hurricane. Rain began to fall; the cabins were crowded, and every part of the ship filled with human beings, principally women. It was such a scene that I did not think I could support it for three hours. The rain, however, ceased, the cabins were quitted by numbers, and I lay down on my back, having twice or thrice had the serious intention of leaving the boat and returning to Scotland. The day, however, improved, the sea became comparatively calm, and we reached Boulogne, after a *beautiful* passage, in two hours eight minutes. An English gentleman and I got a whole carriage to ourselves, in which we went to bed, occasionally getting up to admire the display of summer lightning, which continued for three hours, and was the finest sight of the kind I ever saw.

" When admitted into the Exhibition, I luckily found

there Baron Segur, M. Mathieu, M. Wartmann of Geneva,
and Mr. Tyndall, all particular friends, and the two first
members of the same jury with me in the Crystal Palace
of 1851.   M. Mathieu, the brother-in-law of Arago,
kissed me on both cheeks, according to French custom ;
and Baron Segur, one of the Académiciens Libres of the
Institute, charged himself with getting my tickets. . . .
At three o'clock I attended the first meeting in the
Palais de l'Industrie.   My brother jurors had resolved
to make me president of our jury, which is Class VIII. ;
but the Emperor himself appointed the Maréchal Le
Vaillant, a very able person, to that office, and I am
chosen vice-president.   The affairs of the Exhibition
are managed by the presidents and vice-presidents of
the different classes, presided over by Prince Napoleon,
and we assembled yesterday at three o'clock to settle
some important questions.   The Prince, to whom I was
introduced, presided, and spoke beautifully.   He is *the
very image* of his uncle Napoleon, and corpulent, but a
noble-looking person.   I found at this meeting our
friend Babinet, who inquired after you, and Professor
Willis, Sir William Hooker, Dr. Royle, and several
members of the Academy of Sciences.

"The Exhibition here is *magnificent*, and has been
shamefully traduced.   The Galerie des Machines far
surpasses anything in our old Crystal Palace.   It is
4500 feet long !   With love to Lydia and David, and
kisses to the children, I am, my dearest Maria, your
affectionate father,                    D. BREWSTER."

"PARIS, 1855.
"I have just come from a committee of British
jurors, who have been passing resolutions respecting the

great progress made in France in the mechanical arts, and calling upon the Government for an increased encouragement of science in England. We were very unanimous on the subject. Professors Owen and Wheatstone, Mr. Graham, Master of the Mint, Mr. Rennie, Mr. Fairbairn, Mr. Cockerel, M. Manby, Dr. Hoffman, and the two De la Rives, were among the number. I was the chairman, and got up the meeting."

"PARIS, *July* 1855.

" ON Monday last I spent a very pleasant day at the weekly meeting of the Institute, where I met with many of my distinguished colleagues. Sir John Herschel, who has been for some time dangerously ill, is proposed for the next vacant place among the eight Foreign Associates of the Institute. I shall have the pleasure of voting for him next Monday. I saw there M. Leverrier, who asked me to dine that day at the Observatoire, where I met Dumas, General Morin, and a large party. I handed Madame Leverrier, a very handsome and clever person, to dinner, and was obliged to speak the most awful French you ever heard. She could speak a little English, which was sometimes called for in an emergency.

" I have had a hard day's work in examining microscopes, some of which were magnificent. The more I see of the Exhibition the more I admire it.

" Next to the Pavilion, containing the jewels of the Empress, the most popular objects here are the stands which several opticians have erected, with six, eight, or ten stereoscopes, containing binocular pictures. They are always crowded, but the ignorant spectators are generally seen looking in with only one eye. In passing, I have often astonished them by making them look with both."

This busy summer, his life was still further enlivened by the arrival from India of his eldest surviving son, with his wife and family; he had married in 1849,[1] but his wife, till this return home, was personally unknown,— she soon, however, became warmly beloved and cherished, possessing peculiar affinities with her father-in-law's mind, while for nearly two years the old philosophic house was cheered with the sights and sounds of pattering feet, merry voices, and rosy cheeks of childhood.

During the autumn we went to pay a most congenial visit to Mr. and Mrs. Bateman at Cardross House, in Perthshire, which they then rented; the other guests were Mr.[2] and Mrs. Fairbairn, and Dr. and Mrs. Romney Robinson—he, the celebrated astronomer of Armagh, for many years much esteemed and admired by my father, and she, one of the sisters of the happy family group of Edgeworthstown. The microscope was the favourite subject of experiment and conversation between the two old philosophers during that happy visit, a friendly pond in the neighbourhood furnishing excellent specimens of that wonderful teeming world, into which the microscopic lens is a window through which we may descry the marvels else invisible. One remarkable individual, rolling about like a tiny hippopotamus, was greatly appreciated, and was pronounced to be the *Proteus amœba*, one of the lowest forms of life. Dr. Robinson's anecdotes about the microscope much interested

[1] Married, October 6, 1849, to Lydia Julia, eldest daughter of Henry Blunt, Esq., of Her Majesty's Indian army.

[2] Now Sir William Fairbairn, Bart.

my father. One in particular I jotted down, of an
Italian infidel who scoffed at the Bible allusion to the
" fine linen of Egypt," for he said that there had
been no flax grown in Egypt, and that all the
material found in its ancient tombs was cotton. A
microscopic friend of Dr. Robinson, in examining the
fibres of linen and cotton with the microscope, found
that there was a very decided difference between them;
the cotton being quite flat, with sharp turned-up edges,
which is the reason that it is bad for dressing wounds,—
linen, on the other hand, being composed of round smooth
tubes. The stuff found in the Pyramids and elsewhere
was subjected to this test, and it was found that in
every case it was true " linen," and in many cases of
peculiarly " fine linen." Some of the Indian muslins,
which are the finest in the world, contain 120 threads
in the eighth of an inch, but the old Egyptian linen can
sometimes boast of 130 in that space.

In the late autumn of 1856, Sir David made arrangements which allowed him to spend the winter in the
south of France, the climate of Cannes having been
advised for the health of one of his family. The rest of
the party preceded him, and he arrived in the beginning of December at the Château de Ste. Marguerite,
which had been carefully prepared for his reception.
This lovely villa, which is still called by some " Sir
David Brewster's house," was then unfinished, but notwithstanding was a most comfortable residence. The
views of the Esterels and the Mediterranean, the Iles
des Lérins, the promontory and gulf of Napoule were
lovely—the climate exquisite, the society most congenial, and the neighbouring library and philosophical
instruments of Lord Brougham were at my father's

disposal in his absence, yet nothing seemed to prosper that winter. Climate, friends, scenery, the interests of the Château Eléonore Louise,[1] availed little. He was thoroughly uncomfortable, to the despair of an Italian courier, first-rate cook, and man-housemaid all in one, who would have done anything in the world to please "Seer David," he and a young Niçoise peasant, his aide-de-camp, having both the deepest reverence and admiration for their master, whose scientific eminence had somehow transpired. The mystery was not solved for some time, when it appeared that Nice possessed attractions which could not be found at Cannes. On the 17th of November, Sir David had travelled to Cannes by diligence, with three young English ladies on their way to winter at Nice, on account of delicate health. Many congenial subjects of conversation were found, and the breaking down of the diligence in the Estrelles at midnight, led to greater intimacy. That day's journey was his first introduction to Miss Jane Kirk Purnell, second daughter of Thomas Purnell, Esq., Scarborough. He followed her to Nice in January, and spent a congenial winter in the large English society there. . On March 27, 1857, he was united to this lady,—a marriage which brought a great accession of happiness to his future life; he found in her a most attached and appreciating companion during the years of brilliant life, social and scientific, which yet remained. After their marriage they went by the beautiful Riviera to Leghorn, and thence by sea to Civita Vecchia, arriving at Rome on the 6th of April. It was my father's first visit, and the six weeks of his stay were occupied by an amount of energetic sightseeing which few men of his age could have accom-

[1] Lord Brougham's villa.

plished. He kept an almost equally minute journal as in 1814, looking at every object in its relation to science.

The subject of optical illusions had long been very interesting to him, and especially the class of phenomena which he described in Natural Magic "as that false perception in vision by which we conceive depressions to be elevations, and elevations depressions, or by which intaglios are converted into cameos, and cameos into intaglios."

The following extracts from the journal I put together, as bearing on this curious subject :—

"*April* 29*th, Rome.*—Visited the church of St. Agostino, in which is the celebrated statue of the Madonna, to whom so many votive offerings have been made. She is covered with diamonds and precious stones of all kinds, her fingers with rings, silver lamps lighted are hung before her. The walls are covered with pictures, representing the accidents and events in consequence of which the offerings were made, and hundreds of silver hearts are put up on the pillars in glass-cases, so that the church is like a jeweller's shop. Above the arches there are ornamental paintings, like festoons of flowers, hanging from two pillars ; but in place of appearing in relief, as they ought to do, and as in some positions they actually do, they seem hollow like intaglios. This is the first time that I have seen this illusion in the case of a painting *in plano.*

"*May* 5*th.*—I went again to the church of St. Agostino, and found all the windows in some degree darkened. The optical illusion did not appear, as might have been expected.

"*May* 11*th.*—Went at eleven with Lord Northesk, Mrs. Dennistoun, and Lord Rosehill to the Vatican, to see the MSS. M. Tessier, who was to have shown them, was unwell. We walked through the library and saw better than before the many charming objects it contains. After entering its long gallery at its middle, we went to the right, and saw the curious optical illusion of steps. The light that came in at the side windows, and the dark spaces between, gave the long vista the appearance of steps.

The observer at *AB* saw *AB* reduced to *ab*, and the flat floor appeared to rise to *b* by steps.

" *Thursday, June 4th, Padua.*—The church of Santa Giustina is also beautiful with its many domes. In resting near the main door I was surprised to observe that the nearest of these vaulted roofs, viz., those above our head, appeared extremely shallow, while the next appeared much deeper, and the third deeper still. Every one of our party saw the same effect, and could hardly be convinced that it was an illusion of perspective, which I found it to be by placing myself under the remotest of the three roofs, which then appeared the shallowest.

" *June 8th, Venice.*—I observed to-day an interesting optical illusion when looking at some Terrazzo Scagliola pavement, consisting of a number of black, white, and reddish-brown marbles. With two eyes it seems to be uneven, with slight heights and hollows, the hollows appearing to be where there are fewest black and white pieces. I cannot explain this on the principle of the

chromatic stereoscope. It appears to arise from the
distinctions of the black and white pieces, which makes
us suppose them nearer the eye. . . .

" A brightly painted landscape, in which no relief
appears when viewed with both eyes, has a distinct
semi-relief when viewed by both eyes with a reflection
from a distant mirror. Does this arise from the greatly
increased distance of the reflected picture ? For reflections can hardly produce any effect. I could not place
the picture at a greater distance to try the experiment."

Of some of these observations he made use in a very
interesting paper " On some Optical Illusions connected
with the inversion of Perspective," which was, I believe,
read at a meeting of the British Association, but I give
it here as being probably new to most of my readers.

" One of the most remarkable cases of this kind,
which has not yet been explained, presented itself to
the late Lady Georgiana Wolff, and has been recorded
by her husband, Dr. Wolff. When she was riding on a
sand beach in Egypt, all the footprints of horses appeared
as elevations in place of depressions, in the sand. No
particulars are mentioned in reference to the place of
the sun, or the nature of surrounding objects, to enable
us to form any conjecture respecting the cause of this
phenomenon. Having often tried to see this illusion, I
was some time ago so fortunate as not only to observe it
myself, but to show it to others. In walking along the
west sands of St. Andrews, the footprints both of men
and of horses appeared as elevations. In a short time
they sank into depressions, and subsequently rose into
elevations. The sun was at this time not very far from
the horizon on the right hand, and on the left there
were large waves of the sea breaking into very bright

foam. The only explanation which occurred to me was that the illusion appeared when the observer supposed that the footprints were illuminated with the light of the breakers, and not by the sun.

"Having, however, more recently observed the phenomenon when the sun was very high on the right, and the breakers on the left very distant, and consequently very faint, I could not consider the preceding explanation as well founded.

"Upon attending to the circumstances under which they were now seen, I observed that the human footprints were all covered with dry sand that had been blown into them, so that they were much brighter than the surrounding sand, and the dark side of the impression next the sun; and hence it is probable that they appeared to be nearer the eye than the dark sand in which they were formed, and consequently elevations.

"After repeated examinations of them I found the footprints appeared as elevations as far as the eye could see them, and they were equally visible with one or both eyes. But whenever the eye rested for a little while on the nearest footprint it resumed its natural concavity. I have observed other illusions of this kind which were more easily explained, though they differ from any hitherto described. In the Church of St. Agostino in Rome there is above each arch a painted festoon suspended on two short pillars, but instead of appearing in relief, as the painter intended, by shading the one side of them, they appeared concave, like an intaglio. In other positions in the church they rose into relief. Upon a subsequent visit to the church, I found that the festoon, or suspended wreath, was *concave* when it was illuminated by a window *beneath* it, and in *relief*

S

when the eye saw that it was illuminated by a window above it, the object being similarly illuminated in both cases. In the common cases of inverted perspective, the eye is deceived by looking at the inversion of the shadow in the cameo or intaglio itself; but in the present case the eye is deceived by perceiving that the body painting, supposed to be in relief, is illuminated by a light either above or below it.

"An optical illusion of a different kind presented itself to me in the Church of Santa Giustina at Padua. Upon entering the church we saw three cupolas. The one beneath which we stood appeared very shallow, the next appeared much deeper, and the third deeper still. They were all, however, of the same depth, as we ascertained by placing ourselves under each in succession and observing that it was always the shallowest."

I select a few extracts from the rest of the journal as being in some measure characteristic :—

"*April* 25*th.*—Went with Mrs. Dennistoun, Mrs. Amos, and Lady B. to be presented to Cardinal Antonelli, who resides in the storey above the Pope's apartments in the Vatican. We ascended a magnificent stair of white marble, and saw in passing the large and splendid hall of entrance to the Pope's apartments, in which many of his guards were seated. On entering the Cardinal's apartments we met a gentleman, Count Medici Spada, who was just leaving the Cardinal, and who shook me warmly by the hand, complimenting me about what I had done in mineralogy. The Cardinal was with his friend outside of his own room, and received us at the door with a warmth of manner very unusual in formal presentations, shaking each of us by the hand and expressing his happiness in receiving us. After some general conversation, which was very animated on his part, I said to him that I had heard he had made a collection of minerals. He replied, ' No, I have only made a collection of marbles,' which he proceeded to show us. They were in a beautiful cabinet, each drawer having shallow compartments to receive the speci- mens, which were all of one size, about six inches by four, and half-an-inch thick. In the same cabinet he showed us a beautiful collection of precious stones of all kinds. The uncut diamonds were placed in silver handles or cups in order to take them in the hand to observe their geometrical forms. In a separate box there were eight or ten gorgeous rings, in each of which was a magnificent emerald, sapphire, ruby, or tur- quoise, surrounded with brilliants. The pink oriental topaz, the colour of which was not produced by heat, was the finest I ever saw. The collection of agates was very fine, and also the ancient cameos were finely en- graven. I was much struck with Cardinal Antonelli : a more interesting person I never met with. His looks, his manner, and his intelligence were all of a high order. He was tall, thin, and sallow, dressed in a singular blue cloth dress, like a dressing-gown, with red buttons.

" *Rome, Tuesday, May* 12*th*.—Sat for a crayon like- ness of myself to Mr. Lehmann, at his request. Went with M. Volpicelli to see the photographs of M. Dupuis at the French Military Hospital, of which he is surgeon. He uses citric in place of acetic acid, and thus obtains very fine transparent negatives, in which the half tints are well given. · He employs dry collodion, and his plates may be used a month after they are made. His stereoscopic photos. of the ancient and modern buildings in Rome, a set of which he presented to me, are admirable.

"*Tuesday, May* 19.—Left Rome yesterday. From Montaroni we travelled in the dark to Sienna. About a quarter of an hour after sunset I saw a phenomenon of great beauty. From the sun then below the horizon there emanated five diverging beams of red light, very faint, but still distinctly visible to the height of 30° or 40°. What was very interesting was that the five beams formed angles with the horizon of 30°, 60°, 90°, 60°, and 30°, the middle beam being vertical. This arose from the openings between clouds below the horizon being accidentally at the distance required to produce equidistant radiations. During the journey from Acquapendente I observed the process by which a thunderstorm is produced. When there was not a cloud in the sky, there sprung up a small one several degrees above Monte St. Fuore Amiatta, the loftiest hill in the district, near which is the conical one of St. Soldau. This cloud gradually increased hour after hour, descending upon the mountain as it increased, and finally enveloping the whole and extending to some distance around. The clouds thickened and blackened till we saw the rain falling in torrents, thunder then followed, and the rain reached our carriage and became general for a short time. The sky, however, soon cleared, and the rest of the day was fine. Reached Sienna about half-past nine. The portiera at the gate proposed to examine our luggage, but we remonstrated,—and after an angry discussion with our postilion, a bold and powerful man, the portiera wished us a good journey and allowed us to pass. The faccini followed us to the Hôtel Arme d'Inghilterra and insisted upon the fee, as if our luggage had been examined. At Radicofani I was particularly anxious to avoid the examination of my portmanteau, as it contained a sealed box wrapped in black paper containing plates of dry collodion that were to be used in London and Edinburgh for photographs. They were given me by M. Dupuis, surgeon to the French forces at Rome, and a distinguished photographer. Had the box been opened and exposed to light the plates would have been utterly spoiled. If the officer had seen the box, I could not have made him understand the matter, and he would certainly have supposed it to contain prohibited and valuable articles. Having been told on high authority that every customhouse officer, except Sardinian, could be bribed, I laid down before the officer a piastre. He thanked me very graciously, but returned it, saying it was not necessary.

"*Florence, Friday, May* 22*d.*—Called in the forenoon on Dr. and Mrs. Somerville, and showed them the phenomenon of the radiant spectrum. Along with the doctor I called on my old friend and correspondent the Marquis Capponi, but did not see him, as he had gone to a meeting of the Accademia dell' Etrusca for improving the Italian language. I went to see the interesting Church of Santa Croce, with the tomb of Galileo to the left of the principal entrance, and the tomb of Michael Angelo Buonarotti directly opposite; on the right-hand side are the tombs of Macchiavelli, Dante, and Nobile.

"*Sunday, May* 24*th.*—Went at eleven o'clock to the Scotch Free Church, which meets in Mr. Hannay's house, and met with Mr. Menteith. At three went again to hear Mr. Hannay in the Swiss Church, where the psalms were played on an organ without a human voice. I was here introduced to the celebrated poet Mr.

Robert Browning, who reminded me that I had dined with him several years ago at Judge Talfourd's. Mrs. Browning having just lost her father, was not there. Mr. Browning kindly offered to be of any use to us whilst here. I yesterday met a funeral attended by about six persons in masks. They were gentlemen, members of a Society who engage to take charge of and bury persons killed by accidents, or who have no relations. The Grand Duke is a member, and takes his turn in this office of charity. In the chapel called Misericordia, opposite the great Campanile, the religious services are performed for this class of persons. In an apartment next to it are a number of beds, carried on handspikes, for conveying the persons to church or to their grave.

"*Monday, May* 25*th.*—Went to see the Museum of Natural History and the Observatory. The Museum is, I believe, the finest collection in the world, occupying a vast number of apartments elegantly fitted up. The collection of physiological, botanical, and anatomical preparations is particularly interesting, while those of the mineral and animal kingdoms are hardly inferior in relative importance. I could not discover the order in which the minerals were arranged. The zeolites, in which I took a particular interest, seemed to be widely separated. As Professor Amici was not at the Observatory, I introduced myself to M. Donati, an intelligent and superior man, who showed me the instruments. He has taken photographs of part of the spectrum, and is engaged in examining the lines in the spectra of the stars. I mentioned to him what I had done on the subject, and gave him a slight notice of my experiments on the spectrum.

SIR DAVID BREWSTER.      279

" I then called upon Professor Amici, whom I found in the Via Rennini, in a fine palazzo which he had purchased from Prince Demidoff.  He showed me his large achromatic telescope.  The flint-glass was made by Guinand.  In the Observatory I saw a very fine Newtonian telescope also made by him.

" *Tuesday, May* 2 6*th.*—I met Professor Amici to-day at the Museum of Natural History, in order to examine the telescopes and other instruments which belonged to Galileo, and which are carefully and elegantly preserved in the magnificent tribune erected to his memory by the Grand Duke.  I was permitted by M. Politi, the keeper of the tribune and of the collection of physical instruments, to look through both the telescopes.  One with a plain tube, and as plainly fitted up, was that with which he made his discoveries.  The other was fitted up in a tube of leather, gilt in several places, and was a present from the Grand Duke.  We took both of them to the garden, and looked at the most distant trees and houses.  The aperture of Galileo's own instrument was reduced to about one-third of its area by a diagram made with a piece of card.  The field looked like a small hole, and did not subtend an angle of more than twelve or fifteen minutes.  There was comparatively very little colour, and the vision was very distinct when the object was not luminous. Astrolabes of Alphonso are in the tribune, and many very ancient ones.  Among Galileo's papers was found a correct drawing of the ring of Saturn, but the whole of his body was visible.  I saw also the original drawing made by his son of his escapement, which is very ingenious.  All these facts will appear in the new edition of his works now publishing at Florence, in about twenty volumes. We then examined the cabinet of physical instruments, the finest doubtless in the world ; a collection of an historical kind, exhibiting the progress of invention in all departments of physical science. There is in it a natural loadstone of enormous size with an armature, but the poles are not correctly ascertained. In the tribune of Galileo are preserved all the instruments of the Accademia del Cimento, the brass globes compressed and non-compressed. I then went to the establishment in the Convent of Santa Maria del Novello, and saw their apparatus for making the celebrated perfumes. It is a very fine sight. The shop is filled with little bottles of alkerines, etc., which are most beautifully arranged. There is an elegant apartment with four doors containing mirrors on each of the four sides of the room for the reception of the Grand Duke or his friends when they visit the establishment. The Chiastro Viride, but particularly the Chiastro Grande, are beautiful. In the evening we went to a party at Mrs. Stuart Menteith's, where Messieurs Corridi, Govi, and other professors of the Instituto Technico, and many English, were present. We saw there the beautiful photographs of Mr. Bennets, taken at Naples, etc., with a fine pocket microscope by Amici, brought by M. Govi. We saw Froment's microscopic writing, and I showed with it the remarkable microscopic photographs sent me by Mr. Dancer of Manchester.

" *Wednesday, May* 27.—At eleven o'clock we drove with Mr. Hannay to Arcetri, to see the house of Galileo and the Tower of San Gallo, in which he made his observations ; but no traces of him now remain. . . . The house is a large and ugly villa, with a nice garden about a quarter of a mile from the tower at a lower level, and now belongs to a gentleman in Florence.

" *Saturday, May* 30*th.*— Reached Galileo's house, which is beyond the church, at the right hand, and is indicated by the following inscription above the door, cut in marble :—

> ' QUI OVE ABITO GALILEO
> NOVI SOLEGNO PREGARSTI, ALLER
> POTENZA DEL GENIO LA MAESTA
> DI FERDINANDO II. DEI MEDICI.'

The number of the house is 1600. I went into the garden, but was told that there were no memorials of Galileo in the house. The woman who showed me the house pointed out what she called the meridian line on the back of it.

" *Sunday, May* 31*st.*—After hearing Divine service at Mr. Hannay's, Via Serrago, I drove to Professor Amici's, Palazzo Demidoff, and went with him in his carriage to the Pitti Palace to see the Prince Héréditaire. We found there M. Frescobaldi, the equerry to the Prince, M. Antenoni, his *majordomo*, and M. Simonelli, his tutor. The Prince is rather good-looking, and short, and has very frank and agreeable manners. I showed him, in M. Amici's compound microscope, the microscopic photographs of Mr. Dancer, with which he was surprised and delighted. The Princess also came to see them. She is very beautiful and handsome, but looks in bad health, and not very happy. I spoke to her about her uncle, the late King of Saxony, and Dr. Carus, whom I had met at Taymouth some years ago. M. Antenoni presented me with the interesting quarto volume on Florence which he drew up at the request of the Grand Duke, to be given to foreigners who were introduced to him.

"*Padua, June 5th.*—Went to the curious antique chapel painted by Giotto in 1306, while Dante lodged with him. We then drove to the University, and found Professor Santini, who showed us the great room, lately much ornamented, in which Galileo lectured. In the physical cabinet, which is very fine, and contains many excellent instruments, I met my old correspondent M. Zantedeschi, Professor of Natural Philosophy. Professor Santini kindly took a place in our carriage to show us the ancient and splendid Botanic Gardens, many of the trees in which are 200 or 300 years old. I was introduced to Professor Viziani, who has the charge of it. It is one of the finest establishments in Italy.

" In the physical cabinet of the University we saw the *vertebra* of Galileo, stolen by Dr. Crocchi when the body was exhumed in Florence in 1757. In one of the public streets we saw the tomb of Antenor. Professor Zantedeschi, now blind, called upon me with the different works on physical subjects which he has recently published.

"*Paris, June 22d.*—Called on M. Duboscq and the Abbé Moigno, and went with the former to M. Foucault's to see the new specula for reflecting telescopes, as made by depositing silver from its solutions on concave or convex surfaces of glass. Mr. Power, an Englishman, has a patent for the process, but M. Foucault has undertaken to perfect it. He showed me two telescopes thus made, one about 18 inches focus, and the other in the Imperial Observatory. Both of them gave beautiful vision. After deposition the silver surface is slightly rough or rather not well polished, but it is polished to perfection by a little cotton and a small quantity of rouge.    The film of silver is at first transparent, transmitting green light, but it afterwards becomes opaque when the film is thicker.  M. Fou-cault showed me the whole process from the commence-ment.    A solution of nitrate of silver and alcohol and ammonia forms a brown fluid.    To a hundred parts of this fluid three parts of the essence of cloves is added, and when the concave glass surface is covered with the fluid, the silver is deposited.    In the two telescopes which I saw, the image is received upon a rectangular glass prism.

" M. Foucault told me that a mercurial surface in re-volution, as suggested by Mr. Buchan, an American, for specula, becomes a parabola only at the pole.    He drove me from the Observatory to the Academy of Sciences, where I met with MM. Mathieu, Milne Ed-wards, Babinet, Pelouse, Poinsot, Wertheim, Vernueil, and others.

" After reading the minutes the President adjourned the meeting on account of the death of Baron Thenard.

" *June* 23*d.*—M. Porro having called upon me, I ac-companied him to his Institutio Technomathique, where I saw his great astronomical telescope with which he has discovered a new star in the trapezium of the Nebula of Orion, a discovery confirmed by Padre Secchi at Rome.    This is a fine instrument, and is made with Guinand's flint-glass.    The observer's eye is always at the same place, whatever be the position of the telescope, provision is made for the bending of the tube,—the axis of the eye-piece being always coincident with that of the object-glass, however much the tube is bent.

Mr. Porro's short telescopes are highly interesting, the
tube being shortened by two plain reflectors. They
cost about £5 each, £6 with a micrometer. His methods
of determining the achromatism of object-glasses and
the correctness of the curvature are very ingenious.
He is well instructed in optics, and not sufficiently
appreciated in Paris. He is a Sardinian, and was an
officer of engineers. His machinery for grinding and
polishing large lenses is very ingenious.

" M. Ferrier and M. Ernest Lacan called upon me.
The latter is editor of *Le Journal de la Lumière*, and
invited me to see his magnificent collection of photographs. M. Ferrier presented me with twelve of his
fine binocular slides, taken in the East and in Spain.

"*Paris, Wednesday, June* 24*th.*—I went at ten to M.
Lacan's, who gave me two fine photographs, one of part
of the Louvre, the other an engraving from a photograph
of a sketch by Rosa Bonheur. I called on M. Nièpce
de St. Victor, Commandant of the Louvre, and saw
his beautiful heliographic invention. His photographic
etchings upon steel-plates placed in the camera are
beautiful. His etchings on marble, etc., the lines being
filled up with coloured wax, are interesting; but his
engravings from photographs are singularly fine. I
then went to M. Ferrier, where I saw a room with
shelves, like those of a library, fitted with five or six
thousand of his binocular slides, taken in Italy, Con-
stantinople, etc. He sent me in the evening other
twelve of these very beautiful stereoscopic pictures."

Sir David and Lady Brewster returned to England in
June 1857, and soon after took up their abode at St.
Andrews. During the summer of this year he paid
several visits to the Bridge of Earn, where some of his

family were residing, and also to Whytehouse, near Kirkcaldy, the residence of the late John Fergus, Esq., M.P. for Fifeshire, in whose house he was more entirely at home than anywhere else out of his immediate family circle. In October he wrote :—" I think if you write *a little* every day, or every two days, so as to make it a pleasure and not a toil, you would not suffer from it. I find it a great pleasure to write, if I have my own time for finishing the article, but it is a great pain to be obliged to produce anything at a fixed time. By writing a little, and carefully correcting it, and even writing part of it *twice* when anything striking is to be produced, you would derive much pleasure from it. If I recollect rightly, I think you never made two copies of any of your writings. This is not always necessary, but, as I have already said, it may be well to do it in some parts."

His share of writing-work continued unabated, and his own composition was as careful and vigorous as ever. He still wrote for various periodicals, contributing at different times to Hogg's *Weekly Instructor, Meliora, Good Words,* and others, besides his steady work for the *North British.* He also published, in 1858, *A Treatise on the Kaleidoscope,* in which he gives an interesting account of the gradual development of the invention of that popular instrument, which he appropriately named from the Greek words—καλὸς, *beautiful ;* εἶδος, *a form ;* and σκοπεῖν, *to see.* The first idea of it occurred to him in 1814, in the course of the experiments on the polarization of light by successive reflections between plates of glass, for which he received the Copley Medal. A few months later, during further experiments, the same multiplied images, with the addition of beautiful tints, again struck him. He tells us that " in giving an account of these experiments to M. Biot in March 1815, I remarked to him ' that the succession of splendid colours formed a phenomenon, which I had no doubt would be considered, by every person who saw it to advantage, as one of the most beautiful in optics." It was " in repeating, at a subsequent period, the very beautiful experiments of M. Biot on the action of homogeneous fluids upon polarized light, and in extending them to other fluids which he had not tried," that Brewster was led to discover the leading principles of the kaleidoscope, which he some time after completed, by " giving motion to objects, such as pieces of coloured glass, which were either fixed or placed loosely in a cell at the end of the instrument."

In July 1858 he was attacked by bronchitis, and could neither eat nor sleep, and felt so weak, that he had a strong impression of his near approaching death ; he rallied from it, however, and in the September following attended the meeting of the British Association at Leeds ; after which, change of air at the Bridge of Allan, and pleasant visits to Mr. and Mrs. Bateman at Ferntower, and to Sir William and Lady Adelaide Murray at Ochtertyre, completed his recovery, and before winter he was once more in busy work with eye undimmed and energy unflagging. In July of 1859 he was again in London, whence he wrote to his wife :—" Having had no sleep since Thursday morning, I rose to-day (Saturday) refreshed for a hard day's work. . . . From Mrs. Corkran's I went with her eldest son, a fine boy of fourteen or fifteen, to see the Museum of Patents at Kensington ; and, after

long walking and omnibus driving, I went to Lord
Brougham, whom I found occupied with deputations.
I had, however, a long conference with him on my
claim, which is to commence by sending my memorial to the Treasury. He is to dine to-day at Holland House with Lord Palmerston, and is to speak of
it to him. After I had left him, and was two or three
hundred yards from his house, I heard some person
calling out my name, and on turning round I found
his Lordship hatless, chasing me to call me back to see
some beautiful photographs of Mignet, Biot, and himself. An old beggar-wife came up to him to ask charity,
but he took her by the shoulder and told her to go
about her business, the people in the street laughing at
this singular scene. After I had left his Lordship, and
the door was shut, he again emerged in order to tell
me of a great meeting on negro emancipation which is
to take place on Monday. He asked me to dine with
him to-morrow, that he might get Madame De Bury to
meet me, but I am engaged to dine with Mr. Herrick."

In September he and Lady Brewster went to attend
the British Association, which met that year in Aberdeen. They were guests at Banchory House, which
was always a pleasure; on that occasion Mr. Thomson
entertained the Prince Consort, who presided at the
meeting, and all enjoyed extremely the frank sociability
of that man, "great" in the highest sense of the word.
Sir David wrote the following account of a day at
Balmoral :—

"FASQUE, *Sept.* 1859.

" We left Banchory at five A.M., set off from Aberdeen
at six, and from Banchory-Ternan at seven, in five omni-
buses containing each about twenty-five outside and four-

teen inside, arriving at Balmoral, a distance of thirty-two miles, at one o'clock. About two o'clock, when the Highland games commenced on the lawn, the Queen and the royal party came out to the flower-garden to see them. Soon after this we all rushed to the great hall to a standing lunch, and when this was over, Jane and I went to a large tent prepared for the members of the Association. On our way there, the Queen and Prince Albert came past us, and recognising Jane, to whom H.R.H. had previously spoken, stopped and introduced her to the Queen. All the Highland clans in full dress were present at the games, and when they were over, and we had had coffee in the great hall, we set off for Aberdeen, and arrived there about half-past twelve o'clock. The day was upon the whole good, but at Ballater, about ten miles from Balmoral, it rained heavily, and the wet philosophers were obliged to dry themselves in the royal kitchen!"

While at Banchory House, he received by telegram the tidings that he was appointed Principal of the University of Edinburgh. This intimation caused a severe struggle in his mind between the claims of his *alma mater* on the one hand, and the reluctance with which he contemplated leaving his old St. Andrews home. He did not finally decide on acceptance till October 10th, and shortly after resigned his appointment at St. Andrews. He remained there for a few months longer, however, till after the marriage of his daughter,[1] in the beginning of the year.

[1] Married to John Gordon Cuming Skene, Esq. of Pitlurg, Aberdeenshire, January 6, 1860.

# CHAPTER XVII.

### CHARACTERISTICS.

WHAT's perfect on poor earth ?  Is not the bird
At whose sweet song the forests ache with love
Shorn of all beauty ?  Is the bittern's cry
As merry as the lark's ?  The lark's as soft
As the lost cuckoo's ?  Nay, the lion hath
His fault ; and the elephant (though sage as wisdom)
May grieve he lacks the velvet of the pard.
BARRY CORNWALL.

THERE was never yet philosopher
That could endure the toothache patiently.
SHAKESPEARE.

THAT would be an unsuccessful picture which was all
light and no shadow; that would be an inferior school
of music which dealt only in concords ; that would be
a poor biography which told only the better part and
threw a veil over the rest ; nay, may it not be said that
that would be a poor life which could recount no pro-
gression by antagonism,—no harmony from discord,—no
light shining the brighter out of the darkness,—no falls,
and therefore no risings again,—no temptations and
therefore no victories ?

This certainly could not be said of David Brewster,
for it was not untruthfully affirmed of him that he had
been " a man of war from his youth."   Life was no bed
of roses to him.   Almost every step was trod with a
difficulty, not the less difficult, that it was often entirely
of his own creating ; whilst those that really existed he
made more difficult by a power of magnifying them as

T

by the lens of one of his own powerful microscopes. This exaggeration was not only of feeling but still more of expression. He used the strongest language to express what to other minds would have been a comparatively small trial or event,—the smallest circumstances connected with food, servants, visits, journeys, or such like, were created by a naturally irritable temper and finely strung nerves into serious events, and if the slightest thing went wrong, were commented on in terms so distressed as would have led a stranger to believe that some calamity of unusual magnitude had occurred. In that work to which his practical life was much devoted—the reformation of abuses wherever found—it is easy to see that this habit of feeling and of expression, did not tend to make it an easy or a placid task. During the years of Brewster's connexion with the University of St. Andrews, constant and many were the causes of irritation—the feuds and the lawsuits in which he was engaged. The affairs of the ancient University had undoubtedly fallen into a lax administration, and many of his principal measures were those of wise practical reform. To those behind the scenes it appeared very evident that while in many cases he was right in the main, he was often wrong in his way of carrying out the right thing, and always thoroughly and singularly unconscious of any fault in himself. The strength of expression, the calm stinging terseness of his letters, and the exaggerated views he would take of a slight failure of business habits, did not tend to conciliation. His power of telling sarcasm was indeed very great, —it was a weapon which he too much delighted to use, and which came too easily to his hand when it wielded a pen ; his entire freedom from it in daily life and speech was, however, as remarkable,—I cannot remember hearing him make use of a sarcastic word or expression. On the other hand, if he caused distress and trouble to others, it was but a tithe of what he caused to himself. A troublesome Senatus meeting, or a quarrel with a brother professor caused him a distress, a gloom, a shadow over his life, which those little dreamed of, who saw him bright and genial in society. It was impossible, however, not to see and admire the real placability which mingled with all the vehement and distressed feelings. Few men more strikingly united a capacity for suffering, with a temperament which could forget the depths through which it passed, so that times which seemed fullest of discomfort and trial, looked to him in the retrospect, bright with happiness. When the affair was over and gone, and the thunder-cloud was spent, he could become as intimate and friendly with those who had most deeply wounded him as if nothing had happened, and if a life of battles it was not without victories, as the true and touching words of his last days testify, " I die at peace with all the world."

Whatever the causes, and wherever is due, the principal blame of all the St. Andrews troubles, their rumour went before him to the Scotish metropolis, and much fear was expressed lest there might be a repetition there of the same events ; yet the same parties said, eight years after, " Would that Sir David Brewster could have lived for ever; we shall never see his like again." The minutes of the University Court of Edinburgh thus expressed their grief "for him as one whose warm interest in the University never abated to the last, and who on the many occasions on which he presided over their deliberations, or was associated with them in business, evinced the sagacity of a clear and disciplined intellect, and the courtesy of a kind and Christian gentleman, while each member of it feels that by his death he has lost a valued and respected friend." One professor writes, " I had the happiness of being associated with your father as a member of the Senate during the eight years in which he presided as Principal of the University of Edinburgh, where he formed fresh friendships, and never made an enemy, nor, so far as I know, excited even a passing unkind feeling amongst us. His strong academical sympathies and expressions were of inestimable advantage to the University." A member of the University Court writes, " I almost uniformly differed from him on ' University politics,' and yet his kindness and courtesy were invariable, and I uniformly found him thoroughly tolerant of opposition. I remember, for example, calling at Allerly with Professor Crawford, soon after Dr. Playfair announced his intention of standing for the University representation. I was then doing my utmost to secure the return of the present Lord Advocate, whom I expected to see ousted from his seat for the city of Edinburgh. I expressed my views with considerable warmth, when Sir David quietly said, ' Now, that is a strong statement of the Dean of Faculty's claims, just hear what I have to say for Dr. Playfair.' I adhered to my side, and he never again even alluded to the subject."

It must not be forgotten, however, that Sir David Brewster left St. Andrews reluctantly, carried with him warm and cordial feelings to his colleagues, of whom a more modern school had arisen, and left behind many attached friends and admirers.

Brewster's character was peculiarly liable to mis-

construction from its distinctly dual nature; it was
made up of opposites, and his peculiarly impulsive
temperament and expressions laid him open to the
charge of inconsistency, although he never recognised it
in himself, conscious that he spoke what was consistent
with the point of view whence he took his observations at the time. Accustomed to look at every subject with the critical investigation of the man of science,
he yet united the feelings of the man of impulse, and
he spoke as moved by either habit. Nothing could
show this better than his views and feelings with regard
to clairvoyance and spirit-rapping. Like many Scotch-
men of genius and intellect, he had had a strong lean-
ing to the superstitious from the days of the steeple
vault and the cottage under the apple-tree, balanced,
however, by a scientific mind which required proof and
demonstration for whatever came before it. His own
quaint confession, that he was "afraid of ghosts, though
he did not believe in them," was as near the truth as
possible. Living in an old house, haunted, it was said,
by the learned shade of George Buchanan, in which
certainly the strangest and most unaccountable noises
were frequently heard, his footsteps used sometimes to
perform the transit from his study to his bedroom, in
the dead of night, in double-quick time, and in the
morning he used to confess that sitting up alone had
made him feel quite "eerie." On one of these occasions,
when the flight had been more than usually rapid, he
recounted having distinctly seen the form of the late
Rev. Charles Lyon, then Episcopal clergyman of St.
Andrews, and an attached friend of his own, rising up
pale and grey like a marble bust. He often mentioned
his relief when he found that nothing had occurred to

his friend, and pointed out what a good ghost story had
thus been spoiled. A certain pleasurable excitement
was combined with this "eeriness," and many will
recollect the charm of his ghost stories, recounted with
so much simplicity and earnestness, and *vraisemblance*
of belief, as on one occasion to be rewarded by the per-
plexing compliment of a fair young listener at Ramornie
fainting dead away. On the other hand, he was equally
fond of giving natural and scientific explanations of
ghostly marvels, and used to dwell with great interest
upon the difficulties of evidence in everything connected
with the supernatural, pointing out the unconscious deviations from exact testimony given by persons of undoubted rectitude under the influence of prepossession.
Much of this mingled feeling he carried with him into his
investigations of clairvoyance and its kindred marvels.
He really wished to believe in many wonders to which
his constitution of mind utterly refused credence, and
this feeling, combined with a characteristic courtesy
and wish to please, often misled those into whose pretensions he was most critically examining. On one
occasion, when the exhibition of a lady clairvoyante
moved his companion to an expression of indignant
unbelief, which was declared to be the cause of failure,
his gentleness and courtesy, smoothing away difficulties,
apologizing for the mistakes of the supernatural powers,
and giving every facility for greater success, prevented
the dim-sighted clairvoyante from recognising the equal
but far more philosophical unbelief which was brought
to bear upon her case. He always affirmed that, of the
many cases which had thus come within his ken, he
had never seen anything so wonderful that he could say
it *could* have no natural explanation, though, of a few,

he said frankly that he could neither see nor understand the solution. He latterly took even deeper views of this school of wonders, searching the Scriptures minutely for passages describing the spirits that "peeped and muttered" of old, or those whose "lying wonders" are yet to come, and giving it as his belief that, if modern spiritualism with its manifestations be a truth, it may be a fulfilment of the prophesied work of the evil one and his agents. His views of the important service or suffering or enjoyment of all parts of the spiritual crea- tion were so high, that even in such an aspect of the case, he had special difficulty in believing that spiritual agents were likely to confine their operations to chairs and tables, badly spelt letters, and mawkish sentiments conveyed from the world of awful thought and intelligence.

Perhaps nearly allied to his tendency to the supersti- tious, there was a certain want of self-control, a curious timidity, and a dread of pain, which he used to express with a *naïveté* which was irresistibly amusing. Several of these stories became quite legendary. Mrs. Harford Battersby writes as follows :—

" In illustration of the great philosopher's singular timidity, my father used to tell the following story :— At the time Lord Rosse's telescope was drawing so many scientific men across the channel, he was asked if he were going too,—' Oh no !' he said, ' he was too much afraid of the sea.' My father tried to represent to him what a simple matter it was ; he thought nothing of it himself ; he just went straight to bed on going on board, and awoke on arriving at his destination ; Sir David exclaimed in unaffected horror, ' What ! go to your naked bed[1] in the middle of the ocean ?'

[1] A Scotish expression for going really *into* bed.

"Another favourite story somewhat betrayed the philosopher's want of self-control : he was talking of a severe fit of toothache he had had, and my father asked him, 'What did you do ?' (meaning what remedy had he applied). 'Do ?' said Sir David, 'I just sat and *roared !*'"

He always declined to have recourse to dentistical operations, never having had a tooth drawn—and his answer to any such proposal always was, "What! would you have me part with one of the bones of my body ?"

Although his timidity had the dual element—displayed long before in the Grammar School and playground of Jedburgh—of never "fearing the face of man," he exhibited much of it in connexion with the lower creation. The whole canine race he looked upon as imbued with probable hydrophobia, while cats he declared gave him an electric shock each time one entered the room. A favourite cat having been introduced into the old house, it one day trotted into the forbidden precincts of the philosopher's room—looked straight at him—jumped on his knee—put a paw on each shoulder, and kissed him as distinctly as a cat could. He was so surprised at her audacity, and so touched by her affection, that he quite forgot to feel the electric shock ; his heart was won—from that time they were fast friends, and every morning the cat's breakfast-plate was replenished by his own hands. One day she disappeared, to the unbounded sorrow of her master ; nothing was heard of her for nearly two years, when pussy walked into the house, neither hungry, thirsty, nor footsore—made her way without hesitation to the study—jumped on my father's knee—placed a paw on each shoulder—and kissed him exactly as on the first day! The joy of the reunion was quite touching, although it was never known where she had been during her aberrations; and when, a year or two after, pussy was obliged to be shot, owing to disease produced by over gastronomic indulgence, the distress produced by the event was so great, that, by mutual consent, we never had another favourite.

The humility, true and unfeigned, of Brewster, was so marked a characteristic, that it cannot pass unnoticed here. Those who knew him best and watched him closest, saw it as clearly as those who met him only in society. It pervaded his life; although there was also the dual element which is found in all successful workers, intense consciousness of the powers which he really possessed. None knew better what he had done and could do; none knew better the limits which the Highest had put upon his intellect; "Hither shalt thou come, and no further." Though, like his master Newton, he was well aware of the value and variety of the pebbles he was able to gather on the shore, yet, like Newton, he saw and recognised the far-stretching ocean of knowledge, in which he could but lave his feet, and nought but humility was possible. It is interesting to compare the account of this quality in the mind of the one philosopher written by the pen of the other.

" The modesty of Sir Isaac Newton, in reference to his great discoveries, was not founded on any indifference to the fame which they conferred, or upon any erroneous judgment of their importance to science. The whole of his life proves that he knew his place as a philosopher, and was determined to assert and vindicate his rights. His modesty arose from the depth and extent of his knowledge, which showed him what a small portion of nature he had been able to examine, and how much remained to be explored in the same field in which he had himself laboured. In the magnitude of the com-parison he recognised his own littleness, and a short time before his death, he uttered this memorable senti-ment :—' I do not know what I may appear to the world, but to myself I seem to have been only like a boy playing on the sea-shore, and diverting myself in now and then finding a smoother pebble or a prettier shell than ordinary, whilst the great ocean of truth lay all undiscovered before me.' What a lesson to the vanity and presumption of philosophers,—to those especially who have never even found the smoother pebble or the prettier shell ! What a preparation for the latest inquiries and the last views of the decaying spirit—for those inspired doctrines which alone can throw a light over the dark ocean of undiscovered truth !"

In a letter from Sir Isaac to Mr. Hooke, on some controverted point in science, the biographer puts into italics the following brief and beautiful sentence, so expressive of genuine humility and appreciation of others :—" *If I have seen further, it is by standing on the shoulders of giants.*"

It has been said that there is no connexion between merit and modesty except the letter *m*, but all who have been brought into contact with first-class minds will acknowledge thankfully the profound humility which has been their general law. Though there is no rule without exception, yet even those exceptions might, upon examination, be found deficient in some impor-tant elements of greatness—perhaps too clear-sighted to the pebbles, and too short-sighted for the ocean.

It was not only that Brewster's humility was reverential with regard to that which is highest and beyond, but it was also reverential with regard to that which was around and beneath. His power of knowledge made him ever on the outlook for it in others, and marvellous was the gift he had of drawing it out or creating it.

One, himself the possessor of genial gifts and genius, remarked,—" When I have been with other great men, I go away saying, ' What clever fellows *they* are ;' but when I am with Sir David Brewster, I say, ' What a clever fellow *I* am.' " This jocular testimony had much truth in it, as many of far less intellect will recollect, so strange a gift had he of keeping his own knowledge out of sight, and drawing forth gifts in others, unexpectedly even to themselves. That he enjoyed the honours which his merits had won is very true, but he disliked flattery or any unnecessary allusion to his successes ; and when with fond pride any compliment was repeated to him, the invariable reply was, " Oh, don't tell me any flummery."

One thing was very noticeable in Brewster's mental formation, which is not in itself a rare gift amongst men of success. In Joseph John Gurney's expressive language, he was " a whole man to whatever he did." Whatever subject he was engaged in he made completely his own, and brought everything to bear upon it, becoming quite absorbed in its individuality. The rarity of this power of absorption in his case, was its combination with so much versatility ; and he more than fulfilled Lord Brougham's definition of " a perfectly educated man,—one who knows something of everything, and everything of something." His mind was like his own kaleidoscope, full of countless beautiful bits, all forming into many beautiful wholes. It was the case with lighter subjects as well as those of his own demesne. He rarely read a novel, but when he did so he became quite absorbed in the characters, discussing and criticising them as if each were a living being. One day he picked up *Uncle Tom's Cabin* at a railway station, and was soon absorbed in its perusal. The curious glances at him of the other passengers drew my attention, and I saw that he was in tears, quite unconscious of observation. On arriving at his destination, nothing was done or thought of till he had finished the story. On another occasion, when reading Macaulay's *History of England*, for the purpose of reviewing it, which he afterwards did enthusiastically, he became so completely interwoven with the exciting events, that when my mother one day entered his study, she was astonished by the exclamation, with sparkling eyes and flushed cheeks, " Only think what that villain James has done next !" causing a moment's perplexity as to who the offending individual might be.

At St. Leonard's College one morning I was surprised to find my father, an hour earlier than usual, established at the breakfast table, upon which was his microscope and an extraordinary-looking old volume, sent for from the University library, at an unprecedentedly early hour that morning. It was upon a very unsavoury subject, and it contained engravings unfavourable to breakfast eating, being neither more nor less than a full and particular account of the natural history of the *pediculus* and its congeners ! The night before, in examining microscopically a piece of mica, my father had descried embedded in it some specimens of ancient

and minute insect life, which were new to him, and
which he thought might bear a family likeness to the
figures in the old book. Portraits were at once taken
of the interesting individual, specimens were sent to
friends, amongst others to Miss Mary King (Mrs. Ward),
who also took a likeness, and the result was a very
interesting paper on the " Acari found in mica." His
delight in the discovery, his complete absorption for
the time in the subject, and his eagerness in describing and exhibiting the hideous little mite, stand out
in memory as one instance of that wonderful freshness
and vigour of mind which he brought to bear upon his
work day by day.

His versatility of pursuits and interests, combined
with his extreme accessibility, naturally produced an
immense network of correspondence. The letters that
remain, from all degrees and conditions of people, on
every possible variety of subject, are really a curiosity,
and being mostly answered with care and punctuality,
show what treasures of replies may yet be gathered in.
Letters from working men abound ; many upon the
most abstruse points of science ; some upon mechanical inventions ; others detailing observations of light
and colour ; one writes " because he is haunted by an
idea about a lens ;" another thanks him for " half promises fully performed ;" another for his " inspiring
letter ;" lady authoresses thank him for " a helping
hand ;" most, indeed, abounding in expressions of gratitude. His letters from men eminent in every department of science, literature, rank, peace or war, would
drive an autograph-collector into raptures of covetous
delight. His habits of minute observation were very
remarkable ; a valuable gift in all, and capable of being

highly cultivated, it was in him both natural and acquired to the uttermost. In the walk, at the meal, in
society, in solitude, there was a constant observing
and experimenting upon some common daily circumstance,—the colours and forms of plants, the eye-
balls of fish and other creatures, the habits of gold-fish,
the gambols of mice, abounding in his old house, the
scratching of snail-shells on the window, the jewels
and the tinted ribbons of his lady visitors, the patterns
of wall-papers and carpets, the shadows of carriage-
blinds, the blues and violets of distant mountains, the
formation of the rose-petals, the surfaces of silk, satin,
cotton-wool, swan's-down, etc., were all matters of interest, expanding into higher relation with some scientific truth or discovery. Such a habit of mind was
necessarily accompanied by the intense love of scenery,
which has been already alluded to. We find it breathing through many of his finest compositions ; and such
a passage as the following owes much of its charm to
the reality of the sensations he depicts :—

" Occupying, as we do, a fixed place upon the surface
of the terrestrial ball, treading its verdant plains, survey-
ing its purple-lighted hills, gazing upon its interminable
expanse of waters, and looking upwards to the blue
ether which canopies the whole, the imagination quits
the contemplation of the universe, and ponders over the
mysterious realities around. The chaos, the creation,
the deluge, the earthquake, the volcano, and the thunder-
bolt, press themselves upon our thoughts, and while
they mark the physical history of the past, they fore-
shadow the dreaded convulsions of the future. Associated with our daily interests and fears, and emblazon-
ing in awful relief our relation to the Great Being

that ordained them, we are summoned to their study by the double motive of a temporal and spiritual interest, and of an inborn and rational curiosity. When we stand before the magnificent landscape of hill and dale, of glade and forest, of rill and cataract, with its rich foreground at our feet and its distant horizon on the deep, or on the mountain range tipped with ice, or with fire, the mind reverts to that primæval epoch, when the everlasting hills were upheaved from the ocean, when the crust of the earth was laid down and hardened, when its waters were enchannelled in its riven pavement, when its breast was smoothed and chiselled by the diluvian wave, and when its burning entrails burst from their prison-house, and disclosed the fiery secret of their birth.

" When we turn to the peaceful ocean, expanding its glassy mirror to the sun, embosoming in its dove-like breast the blue vault above, and holding peaceful communion with its verdant or its rocky shores, the mind is carried back to that early period when darkness was over the face of the deep, when the waters were gathered into the hollow of the hand, and when the broken-up fountains of the deep consigned the whole earth with its living occupants to a watery grave. But while we thus linger in thought over the ocean picture, thus placid and serene, we are reminded of the mighty influences which it obeys. Dragged over its coral bed by an agency unseen, and stirred to its depths by the raging tempest, the goddess of peace is transformed into a fury, lashing the very heavens with its breakers, bursting the adamantine barriers which confine it, sweeping away the strongholds of man, and engulfing in its waves the mightiest of his floating bulwarks.

" When on a Sabbath morn the sounds of busy life are hushed, and all nature seems recumbent in sleep, how deathlike is the repose of the elements, yet how brief and ephemeral is its duration ! The zephyr whispers its gentle breathings—the aspen leaf tries to twitter on its stalk—the pulse of the distant waterfall beats with its recurring sound—the howl of the forest forewarns us of the breeze that moves it—the mighty tempest supervenes, cutting down its battalions of vegetable life, whirling into the air the dwellings and defences of man, and dashing the proudest of his war-ships against the ocean cliffs, or sinking them beneath the ocean waves."

Travelling was of itself a joy to him, especially in any new country ; the liveliness of his remarks and the perpetual calls for sympathetic admiration being not always thoroughly appreciated by over-wearied fellow-travellers; nor was it only scenery to which his love and admiring appreciation was devoted—it extended to every work of God. The following beautiful anecdote of this feature of his character has already been given in Sir James Simpson's interesting Address before the Royal Society of Edinburgh :[1]—

" A near connexion, but not a relative, who in former years often lived in his house, and latterly formed one of the three loving watchers by his deathbed, writes me this characteristic and striking anecdote :—' When we were living in his house at St. Andrews twelve years ago, he was much occupied with the microscope ; and, as was his custom, he used to sit up studying it, after the rest of the household had gone to bed. I often crept back into the room on the pretence of having

[1] Now published by the Tract Society.

letters to write or something to finish, but just to watch
him.    After a little he would forget that I was there,
and I have often seen him suddenly throw himself back
in his chair, lift up his hands, and exclaim, " Good God !
Good God ! how marvellous are Thy works."'    Remem-
bering these scenes, I, on Sunday morning (the day
before he died) said to him that it had been given to
him to show forth much of God's great and marvellous
works, and he answered, ' Yes, I found them to be great
and marvellous, and I have felt them to be His.' "
    Associated with these characteristics there was the
early and late love of poetry, and the wish to write it
himself, which has been already noticed.    His ear for
rhyme and rhythm was peculiarly good, though he was
himself conscious of the lack of poetic fire and expression, which are gifts distinct from the deep inner sense
and love of poetry.    His prose, however, was often far
less prosaic than his poetry, and the music and con-
sonance of its stately march show that within him
there were the elements of the true poet.    My earliest
recollection is that of sitting upon his knee while he
read aloud *Gertrude of Wyoming,* his voice faltering and
his eyes filling with tears at the more pathetic passages.
" Sir Walter's " poems were also read aloud in the same
way.    His strong wish was to have those of his family
who manifested any scribbling tendency to turn their
attention entirely to poetical composition, and he therefore wholly discouraged at first any prose attempts.
    His charm in society was great ; he mingled in it
without thought of himself, contented and grateful
because of the universal kindness he received, and ever
on the watch to see and admire something good and
beautiful in others.    One clever old lady who much

U

prized his homage, said in a quaint pet one day, "It's no use to be admired by Sir David,—he admires everybody!" His kindness and love of children were marked. Many of mature age can recall some little act of his thoughtful, cheerful kindness long years ago. One lady tells me what an impression a trifling circumstance made upon her in her childhood. She had asked the philosopher to draw something in her album, he took it to his room, and after many patient attempts brought it the next morning, with a mortified confession that he had completely failed; and then, to make up for this mutual disappointment, he drew a few lines with pen and ink on one page of the album, folded it down, and produced by blotting it the figure of a symmetrical vase,—a simple but pretty experiment, with which I have seen him keep a whole circle of young people in amused and varied occupation. The following playful letter to the daughter of his friend Mr. Lyon, shows the kind way in which he remembered the requests of others ; having been asked to use his influence that a son might be allowed to study art, which was not the profession desired by his father :—

"My dear Jessie,—I had a walk of nearly two hours with your papa yesterday, and after settling a deep theological question, we had a long talk about James. I think I made an impression upon him, so far as to induce him to think seriously of sending him to the School of Design. Your papa thinks that you have too exalted an opinion of James's powers, though he admitted that he had 'a wonderful talent for painting *horses*,'—as if this were all he could do. You must therefore make James turn his hand upon *cows, pigs,*

and *poultry*, and, if possible, empty Noah's ark upon any
of the carpet canvasses you can command.  I would recommend also the more poetical subject of Daniel in the
*Lions' Den*, and if after this you fail in your plan, which
a lady seldom does when she chooses to lay them well
down, I would recommend as James's last resource a
picture of *a herd of Covenanters* on the hill-side, in which
he may place me, in the richest caricature, either of a
deacon or an elder.  Your papa has just been here
asking me to attend a sermon which he is to deliver
to-day at three o'clock, which I of course will do.  I
told him that I was going to write you, without mentioning the subject, so that you must follow up what I
have done.—I am, my dear Jessie, ever most truly
yours,				D. BREWSTER."
  " ST. LEONARD'S COLLEGE,
	*Nov. 5th*, 1850."

A striking dual feature of my father's habits was
the order which prevailed in the midst of apparent confusion.  No "antiquary" more dreaded the advent of a
housemaid or a duster, and yet all his books, papers,
and instruments were in a state of perfect arrangement
and preparedness for his own use, although unintelligble
to others.  His powers of contrivance and "garrin'" [1]
the most unlikely things "do" his bidding was to an
amusing extent.  Much of his apparatus to unlearned
eyes appeared a mass of bits of broken glass, odds and
ends of brass, tin, wire, old bottles, burned corks, and
broken instruments.  Yet it was kaleidoscopic in its
nature, and all resulted in effective and beautiful work.
Experiments in the midst of this dusty medley formed

[1] *To gar* is a Scotch verb for *to make.*  A "gar-doer" is a very old
Scotch expression for any one with the special gift mentioned above.

the chosen and delightful occupation of his life.    Writing was performed " doggedly " as the labour and the duty, but the long dark passages, the round hole or chink in the shutter, the ingeniously cobbled instrument, as well as his more elaborate telescopes and microscopes, formed the material of his greatest earthly enjoyment.    He always on these occasions indulged in a sort of low purring whistle, which though utterly destitute of music, was the sweetest of sounds in the ears of those who loved him, for then it was known that he was entirely free from all *malaise* of mind or body.

Since I wrote the above, I have received a letter[1] which seems to me so descriptive of some " characteristics" of my father, that I make no apology for quoting from it :—

" In anything Sir David had not himself studied he was singularly receptive, making his inquiries with a sort of child-like earnestness that was very touching in one so stored with knowledge.    With all his amazing keenness and subtlety of intellect, and the glancing acuteness with which he would detect any fallacy, it always struck me, notwithstanding, that it seemed to come more natural to him to believe and accept than merely to start objection.    His mind seemed more inclined to belief than to doubt, except in so far as his keenness of vision guarded against anything like credulity.    He was a most patient listener, and was singularly fair and courteous in conversational discussion.    If at any time he started an objection, he was, of all men I ever met, the readiest to admit the full force of anything that might be said in answer.    Sometimes, indeed, as is mentioned also of Goethe, he would take up an argu-

[1] From the Rev. Mr. Cousin.

ment against his own opinions that struck him, repeat
it in his own words, and present it with greater force
and precision. Those who did not know his way, would
sometimes fancy he had accepted their conclusion, when
thus, in the exercise of mere logical clearness and candour, he was but admitting, as he felt it, the weight of an
individual argument. With those less candid or less
logical than himself, his very frankness and candour
in discussion would thus sometimes lead to misconcep-
tion. After such free admission of the force of their
argument, they were surprised to find him afterwards
still retaining his own views."

# CHAPTER XVIII.

## RELIGIOUS HISTORY.

A TA faible raison garde toi de le rendre :
Dieu est fait pour aimer et non pour comprendre.
                                        VOLTAIRE.

THOU know'st my longings to be taught of Thee ;
    All human teaching find I dark and vain ;
Teach me, O Lord, and then shall I be taught
    To know Thyself,—this is my joy and gain.

Unteach me all the errors I have learnt
    In earthly schools ; forgive self-will and pride ;
I would unlearn all falsehood, learn the truth,
    And with Thy truth alone be satisfied.

O Truth of truth ! to Thee, my Lord, I come ;
    Teach me, oh teach, as Thou alone canst do ;
Spirit of truth, come down and fill my soul,
    Fill it with wisdom and with gladness too ;—

The gladness of a glorious certainty,
    Concerning Him who lived and died and rose ;
This, this is true, should all else prove a lie
    And in this truth my spirit finds repose.
                                        BONAR.

" IT is only women and weak men who believe," said
one day a highly intelligent representative of a foreign
nationality. The words seemed to cast a shadow over
the sunny mountain slopes of Lake Leman, where they
were scornfully uttered. " My father, Sir David Brewster, believes entirely," was the reply. The surprise,
the incredulity, the inquisitiveness which this intimation produced, paved the way for a full declaration of
the truths which Brewster's powerful mind had then
fully grasped, and in that and succeeding conversations

there was at least no more said about "weak men."
The last that we heard of our intelligent fellow-traveller
was that he was carefully and candidly reading that
remarkable work, *The Philosophy of the Plan of Salvation*, although with what result I know not. Almost
exactly the same incident happened a year or two later
with another mind of similar tendencies. There came
a day when, in the midst of a joy unspeakable, those two
almost forgotten incidents rushed into my mind, and
I longed to say to those, and many other so-called
philosophic intellects, " Come and see how a Christian
philosopher can die." Doctrines turned into practice—
faith in the future sufficient and strong for the present
—orthodoxy proved to be joy in sorrow and vitality in
death, must ever be the most powerful weapon of the
truth. Sir David Brewster was not one who could or
did speak much with his lips for his Master, but his
full personal acceptance of Christ as the Wisdom of God
and as the accepted Righteousness for man may be of
more use than many sermons to minds cast something
in his own mould. It is in the belief that he being
dead yet has a voice and a message to speak in this
way, that I venture to give a sketch of what I believe to
be his religious history, endeavouring to divest myself
of all personal religious prepossessions. Nothing is
more remarkable at first sight than the decided " orthodoxy" of Brewster's religious opinions. What he called
on his deathbed "the grand old orthodox truths"—the
atonement, the Trinity, election, and the eternity of
punishment,—he held intellectually from the old student
days, when study of the Bible and of the Standards of
the Scotish Church went on uninterrupted by the prosecution of more exciting researches. That this ortho-

doxy in his case was but a barren set of dogmas, giving neither joy, comfort, nor strength, was abundantly evident.    Satisfied with holding them, for years his feet seldom entered the house of God, and his mind was entirely occupied with his absorbing pursuits or carking cares.    From time to time his love of God's works—some literary event or some signal sorrow—turned his closer attention to these things, which he generally thought sufficient to store up in the garners of his intellect. Thus a friend writes :—

"I recollect with what indignation he referred to some unworthy concessions he had found in the letters of Dr. Blair to Mr. Macpherson, in which his religious interest appeared to be merely professional.    I also recollect the dissatisfaction he expressed with an article in the *Edinburgh Review* against evangelical preaching, and especially the doctrine of original sin.    He read to me his correspondence with Professor Macvey Napier, the editor, who attempted to defend what Brewster so effectually demolished."

We have seen also that his son's death awoke much serious thought.    When the claims of a busy life again came before him, it would seem that this temporary interest died away, for the remembrances of speculative childhood all testify to the absence of anything like vitality animating the bones of orthodoxy.

In 1839, when there was a stirring of religious life at St. Andrews, and a steady light kindled in the "city" that used to be called in the old martyr days, "the capital of the kingdom of darkness;" although the movement entered into his family, changing the whole course of life to a young member of it, he gave no sign of interest or participation.    He never opposed or ridiculed beyond expressing his dislike to "extremes," but sympathy he had none to give. At the time of, and subsequent to, the Free Church movement in 1843, of which a sketch has already been given, a deeper interest certainly was manifested in the externals of religion. There was far greater pleasure in "good sermons" and religious topics of the day, while his acquaintance with godly men and learned divines was largely increased. Free Churchism, however, or any other church or creed under the sun, are but as the skeletons of the eastern valley till "clothed upon" by the Spirit of life. After his wife's death in 1850, a great change was visible in Brewster to those who knew him best. It seemed as if there was a breaking up of the fountains of the great deep. His mind and conversation at home constantly turned upon religion. The days of satisfaction with a creed were for ever done with, and it seemed as if he were rent by a great internal struggle. It was characteristic of the dual nature of the mind, that while still clinging tenaciously to the "old orthodox doctrines," so difficult for many minds to receive, he was now beset with cavils upon other points, expressed with the fearlessness of those who have no such restraints of belief, so that he was in that most painful state, a mind and opinions in complete inconsistency, and in a consequent state of civil war and anarchy. Sermons were no longer listened to with interest, weariness, or indifference, but each was fought over in his active mind, till Sunday was indeed anything but a day of rest to him. The principal cavil which beset his mind for a time, was a doubt of the inspiration of God's Word, expressed with a sophistry and ingenuity which brought dismay to those obliged to listen, but utterly unable to reply,—yet this difficulty he never allowed for a moment to intrude into that demesne of orthodox belief which he still proudly kept entire.

The observance of the Sabbath was a subject on which he differed widely from the majority of his countrymen, and on which he could not agree to differ. "Sabbatarianism" always called forth his warmest indignation, and he believed that he had proved (to his own satisfaction) that it was but a shred of Judaism.

Another doctrine which he denounced as a novelty, and which he held in utter abhorrence, was that of "assurance of faith," which he always stigmatized as the height of presumption and self-righteousness. But his principal stumbling-block for many years (and this ought to be rigidly pondered and deeply sympathized with) was the inconsistency of professing Christians. He erroneously held that whenever a man called himself converted, he was thenceforward to be perfect. When any fault or inconsistency, therefore, was visible in his professing friends, clerical or lay, his mind was thrown into a ferment of indignation, and the offender was at once denounced as a hypocrite of the deepest dye. Indeed, so far did he carry this, that he is recorded to have once said confidentially, that he only knew two real Christians. Much of this struggle and conflict appeared to be owing to a sense of something possessed by others, to which he was well aware that, in spite of all his intellect and all his orthodoxy, he had not yet attained, but which he was most certainly seeking, not in the silence awaiting the still small voice, but in the roar of the tempest and the fire of the earthquake. Night after night the Word of God was brought from its place and studied with commentaries and notes, even as in the early days. Whatever pressure of writing, whatever charm of experiment, that large volume was never left to accumulate even a temporary dust. If ever there was a seeking of God with "strong crying and tears," it was by Brewster at this time. Frequently, in the earliest morning, when the writing and the microscope, and the Bible-reading were over, have I been awakened to listen in awe to the sounds of prayer and weeping below.

A year or two later the following touching incident and conversation took place. Mrs. Macpherson writes :—

"It was in March 1856 that I had a long talk with dear Papa upon the suffering of Christ, from which we passed on to speaking of the gratitude due to God. He said he never could feel that there was any such strong ground for a claim of gratitude as people spoke of, since he felt that he had received no more good than was absolutely necessary to enable him to do the work that God required of him. Then we spoke of the possibility of feeling any love towards God, and agreed that such a sentiment of love as is possible between man and man, was impossible between man and God. 'How can we love Him,' he said, 'One whom we have not seen? We admire Him in His works, and trust from the wisdom seen in these that He is wise in all His dealings,— but how can we LOVE Him?'" After this conversation, his daughter-in-law being herself led to understand how alone the love of the unseen Christ can be shed abroad in the heart by the working of the Holy Spirit, felt that she must confess this change in her views and feelings. "He listened most attentively, and when I had finished, took me in his arms, kissed me, and said in such a child-like manner, 'Go now, then, and pray that I may know it too.'"

It was during his severe illness of 1858, that, being from home, I was sent for hurriedly, and found him in such a firm belief that death was at hand, that he had requested his pastor, the Rev. Dr. Ainslie, to pray for him as a dying man. I then first became aware of a change having passed upon him, and first felt assured that there had been some answer to the years of asking, and a time of finding after the long search. During his convalescence it was frequently my pleasant office to read aloud to him. One book, I remember, he listened to with intense interest—Dr. Stevenson on the twenty-second Psalm—the description of the sufferings of our Lord touching him deeply; but the book I remember most vividly, was a small one entitled *Perfect Peace*,— the memoir of a clever and scientific medical man, who accepted Christ after a severe struggle. He listened with his peculiar habit of vivid interest, combined at first with unqualified approbation. In the course of reading, however, we came to several of the biographer's laudatory remarks, as the following :[1]—" *Notwithstanding* his high talents and great proficiency in professional and scientific knowledge, he talks with me in the most child-like manner on the things that concern his peace; indeed he evinces as humble and teachable a spirit as I have ever met with ;"—" his humble and teachable spirit, *notwithstanding* his great literary and scientific attainments," etc. etc.,—" the pride of intellect was cast at the foot of the Cross of Christ." These passages produced evident dissent, and when I read the last quoted, it caused an unpre-

[1] The italics are my own.

meditated burst of disapprobation, more satisfactory
than any set expressions could have been, and though
vehement, was very different from his former bitter
denunciations. "That disgusts me !" was his sudden and
lively exclamation, which made me look up in some
dismay.   "A merit for a man to bow his intellect to
the Cross ! Why, what can the highest intellect on earth
do but bow to God's word and God's mind thankfully ;"
and, he added, with a touching simplicity, "That's not
*my* difficulty—what distresses me so much is that I
don't love the Bible enough.  I have it at my fingers'
ends, but I get tired of reading the passages I know so
well."   The barrier of reserve was partially broken
down, and he went on to say that he had had no fear
in the thought of dying which had been present with
him for several weeks, for that he had entire personal
confidence in the work of his Saviour.   It is interesting
in this connexion to note one of his expressions on his
deathbed.  Speaking of the atonement, he said,—" It
gives me 'perfect peace' in resting on it now."

There was evidently much still wanting—much that
one could neither define nor understand.  But from
that time, for four years, there was much earnest
humility in seeking further knowledge, of which I heard
afterwards from her most with him, and with whom
he had no reserve.   They conversed long and frequently
on "the things of the kingdom ;"—his anxiety for her
full acceptance of what he himself was holding with
a real, but not a joyous grasp, was very touching, and
was often expressed by his use of the little word "we"
in speaking of this mutual search, even after he had
sought and found.   His own earnest cry and need still
was to "love" as well as to "know,"—the difficulty of

feeling love to the Unseen One, being his difficult experience, though he now knew it to be attainable.

In 1862 my father made the acquaintance of Mrs. Barbour, author of *The Way Home*, *The Soul-Gatherer*, etc., whose fine qualities of mind and unreserved dedication of speech and talent to her Lord, had a special charm for him. One day he went to see her, and in his humble childlike way he alluded to some doubts and difficulties, saying, "We are trying to look to the Lord Jesus, but are not succeeding as we would like." A simple illustration of "Faith" and "Substitution," used by her, touched him deeply. Some months after, they met again, and Mrs. Barbour writes :—

"Almost the first word was, 'Are you looking to Jesus?' 'Oh yes,' he said, 'and I see Him plain.' 'And are there no doubts remaining?' 'None,' he answered; 'unless it be sometimes from the entireness of faith. Hitherto it has always seemed as if the truth were a moving sea beneath me, now it is rock; and no wonder if a doubt sometimes cross me whether what seems so natural and so unbroken can be divine.' We read Hebrews xii. 22, to the end. At another time, speaking of the danger and folly of not using God's way of redemption fully, he said, 'The folly is trying to toil up the hill when God has sent a locomotive down for us; and then in our own way who would ever scale it?' That day in November he was rejoicing in Christ as the Healer. I said, 'Think of any one coming from the antipodes to consult a physician, and then trying for a week to make himself better before sending for the doctor.' 'Height of folly!' was his short answer; 'go at once.'"

Some years afterwards, when a poem of this lady's

was read to him, he said, with much emotion, "Dear
Mrs. Barbour! how I do love her! She was once very
useful to me; I owe her a great deal;"—referring,
there is no doubt, to the preceding circumstances. The
thought of the hill Difficulty expressed above was a
favourite one of my father's. Mr. James Balfour remembers that he joined with great animation in a conversation one evening at Mr. Barbour's house, where a
lady stated that she had always thought religion consisted
in "doing her best," and that she had "always kept
rolling the stone up the hill." "Yes!" said Sir David
quickly, "but you would never get it any farther up!"

A few days after my father's interview with Mrs.
Barbour he went to London, whence he wrote to Lady
Brewster:—

"LONDON, *May* 27, 1862.

"MY DEAREST JEANIE,—... I have been very thoughtful about the great subject which interests us so much.
The Exhibition is suggestive of good. The dazzling
display of the wonderful materials within and without
the earth He has created for our use and enjoyment,
is proof of His love and kindness. To convert those
raw materials into the splendid fabrics and structures
which fill this building, He has endowed man with all
the various powers and capacities which were required.
In the exercise of these powers, and in carrying on the
great purposes of His kingdom, we have entirely forgotten the Giver of all these materials, and of all those
varieties of intelligences, and we have ever made them
the ministers of sin, using them in the violation of His
laws, and in injuring our fellow-creatures. Thus exposed to punishment, He has sent His Son to ransom
us, and yet we cannot truly believe that God has done

all these things for us, and find difficulty to accept of the great offer of restoring us to His favour. If these views are kept not only daily but hourly before our minds, and if we anxiously pray for help to realize them, this persistency and anxiety for Divine light must itself yield some peace, which must sooner or later grow into that genuine peace which we are struggling to reach.

"If we are sincere in our search, the very conviction of that sincerity, and the continued anxiety with which we seek it, must surely lead to its ultimate attainment, that is, first to that peace which must be the fruit of striving to obtain it, and then of that higher peace which is the gift of God. I am ashamed to write you so stupidly on this subject, but I am very fatigued, both bodily and mentally, with the excitement and work of the day."

" ATHENÆUM, *May* 30, 1862.

" I do not find, what I was afraid of finding, that the gaiety and bustle of London at all interfere with my serious life. They rather favour it than otherwise. It was impossible to see the gay crowd of thousands of immortal beings fluttering like butterflies from flower to flower in the Exhibition yesterday, without feeling that few of them were deriving any spiritual good from the sight."

" LONDON, *June* 1, *Sunday*, 1862.

" Having been asked to breakfast with Babbage to-day, I could not refuse, as he is engaged every other day at the Exhibition. I was afraid it would prevent me from getting to church, having previously resolved to go to Mr. Chalmers's in Upper George Street, but I tore myself away from Babbage in time to get to the chapel, about three-quarters of a mile distant; you may guess my surprise and delight when I found that Brownlow North was to preach. He gave a powerful and instructive sermon to a crowded congregation. I had the pleasure of shaking hands with Lord Breadalbane, who was there, and who was much pleased with the sermon. Afterwards I returned and lunched with Babbage.

"I dined yesterday at the club here with Sir W. Snow Harris, David Roberts the great painter, Solomon Hart, a Royal Academician, and Mr. Munro of Novar, and was detained a little later than half-past seven; but I had a sweet recollection of your and Lydia's occupation, and I trust was not less importunate in seeking a blessing upon you both."

From this time the voice of the trumpet was ever louder, and the light of faith clearer and yet more clear. Like the Eastern philosophers he rejoiced, he worshipped, he offered the gold, the frankincense, and the myrrh of his mighty intellect. To those who had watched the long struggle, the mental change was great. His former critical judgments of Christian workers had completely passed away; without stint or question, he was ready to accept all who professed to love the Lord Jesus in sincerity and in truth; and most touching it was to hear him use the very arguments which had formerly been pressed upon himself, to modify the hard judgment of others. His interest in work for the good of souls was now warm and unreserved. This feature was thus commented on after his death :—" When first it was my privilege to be brought into personal acquaintance with him, the Churches were all deeply moved

x

and stirred with the pulsations of reviving life from God; and I remember my delight in noting with what deep interest the illustrious philosopher watched the movement, and rejoiced in the accounts of the revival in the Church, and conversion in the individual soul, which were coming from all parts of the country. It was a subject on which he often spoke, and always with the deepest interest."[1]    Another singularly marked change was his unusually full and free acceptance of the long denounced doctrine of "assurance," although he was ever careful to discriminate between the use and the abuse of it.    One cherished remembrance is of a sweet communion Sabbath in a country church in Aberdeenshire.    After the simple service, some delay having occurred with the carriage, we two rested in a little wayside cottage, and there out of the abundance of the childlike heart the mouth of the philosopher spoke.    One very dear to him having been deterred from joining the service by doubts of personal acceptance, he mourned over this, saying with much simplicity, " I see it all so clearly myself. It can't be presumption, to be SURE, because it is CHRIST'S work, not ours; on the contrary, it is presumption to doubt His word and His work."    He then spoke of the finished and complete nature of salvation, and of the " LORD ' being " our RIGHTEOUSNESS," a thought which was to the end peculiarly precious to him.    In 1865, when paying a visit to Mr. Macdowall Grant of Arndilly, at his lovely residence on the banks of the Spey,—sometimes fondly called by friends the " Palace Beautiful,"— which my father enjoyed intensely, Mr. Grant was much struck by his simple views of personal forgiveness, which he expressed on one occasion with force and

[1] The Rev. Mr. Cousin.

clearness on hearing a lady argue this subject with their host.

Brewster now knew, and put into practice, true " breadth," that much-abused word. He had passed through the furnace of doubts and cavils. He knew the difficulties of a consistent walk as well as of a pure doctrine, and there was ever after a tenderness of judgment towards those whom he believed in error on either side. A grave shake of the head was sometimes the only way in which he showed his opinion. He could now be " broad " even to those whose views he still held to be " narrow "—perhaps one of the most difficult phases of catholicity. He never, so far as I know, gave up his own strong views against Sabbatarianism as a doctrine, but he no longer denounced it. He felt the utmost respect for those whose opinions he still differed from, and he acknowledged the inexpediency, though not the unlawfulness, of the measures which would reduce England to a continental Sabbath level.

The old orthodoxy on other points was unchanged, save for the new life that pervaded it. We are told that his mind was clear and decided on all the questions of the day. " He would not unfrequently speak of the supposed contradictions between science and Scripture. He would never admit that there was any real contradiction; and he was alike impatient of the dogmatism of theologians denouncing science, and of the dogmatism of scientific men denouncing theology, as being equally and on both sides founded on error of each other's views. As a man of science he was jealous for his order; and I remember with what a kind of proud satisfaction he put into my hands a list of scientific men of high standing who had avowed their faith

in Scripture.    He referred to this as a token that there
was no natural tendency in science to shake the faith
of men in the Word of God, and no justice in regarding
men of real science as more inclined to infidelity than
others."[1]

A well-known professor of divinity, personally unacquainted with Sir David Brewster, expressed his
earnest wish that the latter would write a short pamphlet
upon the errors of the day, with a declaration of his
own views.    On this application being made known to
him, he wrote to me as follows:—

" It is difficult, if not impossible, to reconcile certain
statements in Scripture with what is accepted by many
persons as science, and imperfect and unsuccessful
attempts to do this are more injurious than beneficial to
religion.    The only mode of dealing with this matter is
to show that the science which is opposed to Scripture
is not truth, and this has been done in a little work, *On
the Age of Man*, by Professor Kirk, in reply to Sir C.
Lyell's speculations on that subject ; and papers of a
similar kind have been published by the Victoria
Institute, which is presided over by Lord Shaftesbury.
In my sketch of Dr. Greville's Life, which I enclose,
I have stated my views on this subject, and also in a
short note to Professor Kirk, which he has published.
As a matter of science, the subject is quite out of my
sphere ; and even if it were not, I do not think I could
do any good by publishing a popular tract upon it."

The following is the extract which was enclosed :—

" Dr. Greville was a man of rare accomplishments,
and of no ordinary virtues.    His studies were not confined to science, nor his ambition limited to the honours
which it merits, and the fame which it brings.    His

[1] Rev. Mr. Cousin.

large heart embraced every measure of philanthropy, whether national or local,—whether originating in vicious legislation, in the necessary inequalities of social life, or bearing upon the victims of ignorance, intemperance, and crime. Nor did he take a less interest in those higher questions which disquiet the inner and nobler life of man, stretching beyond that bourne from which no traveller returns, and affecting interests which time cannot measure nor space define. He had pondered in the lesser world over mysteries which he failed to comprehend,—over marvels of life which startled science and rebuked reason; and he submitted with the same reverence to those deeper mysteries which human instruments had equally revealed. Faith, and science falsely so called, had to him, and has, I hope, to many of us, two opposite horizons—the one where the sun rises, and the other where he sets. In the auroral and meridian light of the one he studied, lived, and died; and in the murky twilight and midnight darkness of the other, he wept over the fallen stars of science,—the Sappers and Miners of our Faith."

Although not sympathizing with Professor Kirk in some of his theological views, my father much admired his forcible little volume and its clever arguments against the prepossessed theories, and the "taking for granted" reasonings of the opposite school. He recommended it to my husband as a useful book to put into the hands of young men who might be wavering in their belief, owing to these discussions.

In a short biography of his friend Michael Faraday given in one of his Presidential addresses to the Royal Society of Edinburgh, he thus gives expression to his own sentiments on these subjects:—" With a judgment thus sound and thus patiently exercised, he had no difficulty in answering the question 'What is truth?' among the complex laws of the material world, and he had none in answering the question as put by the Roman Governor. Like Newton, the greatest of his predecessors, he was a humble Christian, with the simplicity of apostolic times. Among the grand truths which he had studied and made known, he had found none out of harmony with his faith; and from the very depths of science he has proclaimed to the Sciolists, that there is a wisdom which is not 'foolishness with God.'"

And he closes another of these addresses in the following words :—" In the study of nature there is no forbidden ground. Into its deepest mysteries we are invited to dive, and if we make Reason our guide, and Imagination our footstool, we may rest assured that truths that are demonstrated will never rush into collision with truths that are revealed."

With the doctrines of Ritualism he had no sympathy; —much attached as he was to many members of the Church of England, and able also to worship occasionally in her congregations with pleasure and profit, he had yet a strong conviction of her perils. In his will he inserted a clause requesting that his little daughter might be " brought up as a Presbyterian, as the best security against her becoming a Roman Catholic." He had also a marked preference for the simple worship of his earlier days, greatly preferring it for his own use to a Liturgical form. The fact that he never had family worship has often been much commented upon, and at one time it was made the foundation of an unsuccessful and party attempt to deprive him of his eldership in the Free Church. Early nervous difficulties in extempore prayer, which never forsook him, were the cause he assigned, along with his being unable conscientiously to adopt in his family a form of read prayer. This difficulty, which never left him, was much to be regretted, though during the later years of his life he gladly availed himself of the services of those he considered competent for the duty, when staying with him, and always punctually attended.

I feel that I cannot better close this chapter, which I have striven to render as simply truthful as possible, than by the following little record of some of his later words respecting his views of Scripture and its fulness, sent to me by a near neighbour and the daughter of a very old friend of his, with whom he was on most intimate terms during the last eight years of his life :[1]—

" One day dear Sir David, in speaking of the lax views now so prevalent, mourned over them and said, ' As for me, I have the Bible, and it all seems so clear and simple that I find all I want in it—the pure Word of God,' or words to that effect. Another time he spoke with deep regret of a conversation he had had with a person holding very broad church opinions. Tears filled his eyes while he earnestly said, ' Oh, is it not sad that all are not contented with the beautiful simple plan of God's salvation—Jesus Christ only—who has done all for us,' and expressed his feeling that if the atonement in its entireness were to be touched, all this glorious gospel scheme must fall to the ground. How vividly I remembered all this, when, on that Saturday afternoon preceding his death, I saw him quietly waiting for the call, and in answer to some remark of mine he said, ' Ah, yes, what should I have done *now* had I to find a Saviour at this time ? ' "

[1] Mrs. Maconochie of Gattonside House.

# CHAPTER XIX.

## NOTES OF LIFE FROM 1860 TO 1864.

How much is changed of what I see,
  How much more changed am I,
And yet how much is left—to me
  How is the distant nigh !

The walks are overgrown and wild,
  The pavement flags are green,
But I am once again a child,
  I am what I have been.

The sounds that round about me rise
  Are what none other hears,
I see what meets no other eyes
  Though mine are dim with tears.

I mourn not the less manly part
  Of life to leave behind,
My loss is but the lighter heart,
  My gain the graver mind.
                         HENRY TAYLOR.

THIS last and incompleted decade of Brewster's life
began with an uprooting from the old ground, which
was no easy task for one of his great age.  The merely
physical discomforts were much increased by an un-
toward event, with which all who understand the
mysteries of a philosopher's sanctum, will intensely
sympathize.  Upon finally leaving St. Andrews, he
packed his carriage not only with his valuable plate,
but with invaluable papers, and the treasured odds
and ends of his experiment room.  Through the care-
lessness of officials, it was allowed to drop into the
Firth of Forth, in the process of being transferred
from the landing to the steamer.  Although he received

compensation for the damaged plate, yet the injury
to his papers, of which some were destroyed and
others much defaced, nothing could compensate. It
is seldom that such a complete uprooting so late in
life is attended with such a thorough taking root in the
new soil,—new, however, it could scarcely be called, and
he soon felt the delight growing and increasing of the
return, under altered circumstances, to his old University,—the reviving of old friendships,—and the forming
of new. The difficulty of finding a house which would
be suitable for the double purpose of living and experimenting in, led to a measure which greatly increased
the happiness and comfort of his remaining years. He
made arrangements to return again to his old home
on the banks of the Tweed, which was within two
hours by railway of the University. Every winter he
moved into Edinburgh, taking a house in the town or
neighbourhood for three or four months, but during the
rest of the session he attended the meetings of the
Senatus, or any other connected with his academical
duties, going out and in from Melrose generally the
same day, with a punctuality and alacrity which was
indeed marvellous in a man of his age, a habit which
was continued till within a month or two of his death.

After leaving St. Andrews in February, he and Lady
Brewster occupied Strathavon Lodge, a pretty marine
residence at Trinity, three miles from Edinburgh, kindly
lent to them by Sir James Simpson, and they moved to
Allerly in September of the same year.

He was at once appointed Vice-Chancellor, and in
that capacity presided at the installation of Lord
Brougham as Chancellor of the University. It was an
interesting sight to see the two Edinburgh students, as

octogenarians, thus standing together on the platform intimately associated in the highest honours of their *alma mater*.   Never had my father's pale spiritual face and venerable form shown to more advantage than on that occasion, aided, as feminine admirers did not fail to whisper, by the flowing purple of his new robes of office. When in London in the summer of this year, he made, I believe, his first acquaintance with another of those "circuits," in which he became still more deeply interested—the National Association of Social Science. Sir David became one of the Council, and was afterwards chosen a Vice-President.   In September of this year also the British Association met at Glasgow, which he attended ; the following extracts refer to these two meetings :—

"ATHENÆUM, *July* 17, 1860.

"MY DEAREST JEANIE,—Having only half-an-hour at my disposal, I sit down to devote it to you.   There is sitting here at present a great International Congress, with deputies from every part of the world, grouped in six or seven sections, and discussing the most important and interesting practical subjects.   Prince Albert opened it yesterday with a noble address, which was most enthusiastically received.   Lord Brougham presides over the Judicial section, Lord Shaftesbury over the Sanitary one, Van de Weyer, the Belgian ambassador, over another, and so on.   I have to read a paper in the sanitary section on my method of illuminating houses in dark and narrow streets.   I have been running through the different sections this forenoon, and have met with many interesting foreigners and old acquaintances."

"129 BATH STREET, GLASGOW,
*Sept.* 27, 1860.

"I accompany Lord Brougham everywhere, sticking as closely to him as a Vice-Chancellor does to the Chancellor, Sir James Campbell completing the *trio*. His carriage with white horses is known to the Glasgow world, and wherever it is seen crowds follow to see the great man. It was very considerate in you to send me Professor Forbes's letter. It was delivered to me in the carriage with Lord Brougham, and after reading it I handed it to his Lordship, who immediately promised to speak to the Lord Advocate, as suggested in the letter.

"We were on our way to see the great war frigate, 600 feet long, and 6000 tons burthen, which is to carry 50 of the largest guns. It is built of iron, and when we approached the yard the hammering of 1400 men upon iron rivets, joining plates of iron of enormous thickness, was almost deafening. When we entered the ship all the workmen left it, and stationed themselves in the yard in one living mass, cheering Lord Brougham in the most enthusiastic manner, while some hundred boys belonging to the establishment placed themselves on a huge pile of wood, and added their shrill notes to the graver music. The same enthusiasm was shown when we left the ship,—Mr. Napier, the great and wealthy shipbuilder, and his two sons, having accompanied us through the works. . . .

"Two very interesting papers were read to-day, one on the repression of crime, by Mr. Arthur Kinnaird, and the other on education, by Dr. Tulloch, which was excellent and highly appreciated."

"Glasgow, *Sept.* 28, 1860.

" . . . At eight o'clock we adjourned to the City Hall, where Lord Brougham had to address 4000 or 5000 of the working classes, and where resolutions for their approval were to be proposed.

" Such a magnificent sight I never saw, of fine-looking and well-dressed men. The resolutions were in every case seconded by working men, with a power and even eloquence which surprised the gentlemen on the platform. There were three interesting foreigners present, M. Garnier Pages, who was at the head of the Revolutionary Government of France in 1848, M. Desmarets, a celebrated French advocate, and Louis Blanc, who was also one of the Ministers of 1848. The two first made eloquent speeches, Desmarets in English, and Pages in French, every sentence of which was translated by Mr. Arthur Kinnaird. Louis Blanc was not asked to speak, lest he should be indiscreet towards Napoleon. In the other speeches every allusion to the despotism of the Imperial Government was loudly cheered."

In 1861 an event happened which brought a new sunshine into the " old man's home." On the 27th January a little daughter, Constance Marion, was born, on whom he doted with the tenderest affection. Strangely touching it was to see the flaxen hair and wide brow of the little one resting on the silver locks of the venerable head, and as years passed on it was not difficult to discover the promises of an inherited talent, which it was sad to know it was impossible could be watched over and fostered by her distinguished father. In very early years, however, he gave her lessons in astronomy, in drawing, and in arithmetic,— she was the constant

companion of his drives, and recipient of that admiration for the beauties of nature which still in him
seemed ever on the increase. This new possession was
not, however, one of unmixed joy; some of the unphilosophic tendencies of the philosopher, which we have
noticed, came into full operation in everything connected with his little darling. His excessive timidity
about illness, infection, and accident, caused him many
an anxious hour, and the almost forgotten distresses of
colds and "great-coats," alluded to by Miss Edgeworth
in one of her lively letters, came back with a forcible
reassertion, which would have been amusing had it not
mingled a real suffering with the joy brought by the
little sunbeam.

A University deputation took him to London in
March, and the British Association to Manchester in
September. He wrote :—

"ATHENÆUM, *March* 3, 1861.

" MY DEAREST JEANIE,— . . . At an interview with
Mr. Disraeli yesterday, I was the last of about twenty
that came into the room, and having been announced
by name, Disraeli walked half-way up his long drawing-room, shook hands with me, and said that it was a
long time since he had the pleasure of meeting with
me. I had utterly forgotten having ever met him,
but I begin to remember that Mr. Lockhart brought
him one day to Allerly when he was a very young
man, and on a visit to him at Chiefswood."

"THE POLYGON, MANCHESTER,
*Sept.* 4, 1861.

" I arrived here last night, and found that my *home*
was to be at Mr. Hilton's, next door to Mr. Fairbairn's,
where Sir R. Murchison is also to sleep and breakfast.

. . . The party at Mr. Fairbairn's is charming, the Harcourts, Romney Robinsons, Sabines, and Batemans. . . . Dear little Connie and her dear mother are never out of my mind, and in my mind's eye I see her clinching the side of her bath with that charming intelligent smile, which I see as distinctly as if I were beside her.

"*Sept.* 8.—We had a very interesting evening meeting yesterday after dining with the Fairbairns. Professor Grove gave half-an-hour's lecture on the telegraph, nearly seventy forms of which were displayed at the meeting. Arrangements were made to receive messages and return answers from Balmoral, Petersburgh, and Odessa, the telegraph wires being brought into the hall where we were. The Prince asked if the meeting was successful, to which we returned the number of members here. We learned from St. Petersburgh, in answer to inquiries about the weather, that the night was fine and the thermometer at 17° Réaumur (70° Fahr.), and we learned from Odessa, round by St. Petersburgh, that the night was cold and windy."

In 1860 Sir David Brewster received an honour which he considered one of the most gratifying which he had ever received. He was made an M.D. of the University of Berlin, and the intimation was couched in the following terms :—

" Die medicinische Facultät hiesiger Universität hat bei Veranlassung des 50 jährigen Gründungsfestes der Universität Ihnen der grossen Verdienste um die Hülfswissenschaften der Medicin halber, welche Sie Sich mit allgemeinster Anerkennung erworben haben, den Grad eines Ehren-doctors der Medicin bei der hiesiger Universität zuerkannt, und diess am $16^{nten}$ in feierlicher Sitzung im Beisein der hohen Staatsbehörden öffentlich erklärt."[1]

One practical application of these "auxiliary sciences" was the beneficial suggestions as to the cause and cure of cataract, which were the result of Brewster's optical investigations. He mentions, in a paper " On the Cause and Cure of Cataract,"[2] that, about the year 1825, he himself had had an incipient threatening of that complaint, which he first became aware of when playing at chess with Sir James Hall, who was "a very slow player." In this situation the active mind occupied itself with experimenting upon the flame of a candle, when he became aware of a luminous and partially coloured halo around the flame, and also, as subsequently seen, around the moon and other centres of radiance. This affection lasted for about eight months, causing him, like the other threatenings of his eyesight, the greatest anxiety. His previous examination of the eyes of animals, especially those of the sheep, the cow, and the horse after death, made him discover the cause of this unpleasant phenomenon, which was a separation of the laminæ of the crystalline lens and a partial drying up of the albuminous fluid. His attention was thenceforward much directed to the study of this subject, which he discussed with many medical men, and he stated at the meetings of the British Association of

[1] " The Medical Faculty of the University of this place has, on the occasion of the Half-century University Foundation Festival, awarded to you the degree of Honorary Doctor of Medicine, on account of the great services which you are universally recognised to have rendered to the sciences auxiliary to Medicine, and this has been publicly declared on the 16th instant in a solemn assembly, and in presence of the high officers of Government."

[2] Printed in the *Transactions of the Royal Society of Edinburgh*, 1865.

1836-37, his conviction of the effectual cure of incipient cataract in two ways, which have since, I believe, been frequently acted upon.

"1*st*, By discharging a portion of the aqueous humour, in the hope that the fresh secretion, by which the loss is repaired, may contain less albumen, and counteract the desiccation of the lens.     2*d*, By injecting distilled water into the aqueous chamber to supply the quantity of humour discharged from it."     The first of these methods was suggested to him by the examination of that "case of conical cornea," and its surgical treatment, which he briefly recorded in Kearsley's Ledger-Book—nothing that he ever saw or "examined" being lost on his retentive mind.     Another benefit to the Art of Healing which accrued from his studies on Light was his practical application of it to sanitary requirements. The following statements of his views as to the lighting of the poor man's home are very interesting :—

"In treating of the influence of light as a sanitary agent, we enter upon a subject almost entirely new ; but admitting the existence of the influence itself, as partially established by analogy and observation, and asserting the vast importance of the subject in its social aspects, we venture to say that science furnishes us with principles and methods by which the light of day may be thrown into apartments which a sunbeam has never reached, and where the poisons and the malaria of darkness have been undermining sound constitutions, and carrying thousands prematurely to the grave. . . . Could we investigate the history of dungeon life, of those noble martyrs whom ecclesiastical or political tyranny have immured in darkness, or of those felons whom law and justice have driven from society, we should find

many examples of the terrible effects which have been
engendered by the exclusion of those influences which
are necessary for the nutrition and development of the
lower animals. . . . If light develops in certain races
the perfect type of the adult who has grown under its
influence, we can hardly avoid the conclusion drawn by
Dr. Edwards, 'that the want of sufficient light must
constitute one of the external causes which produce
those deviations in form which are observed in children
affected with scrofula,'—an opinion supported by the
fact that this disease is most prevalent in poor children
living in confined and dark streets. . . . The problem
which science pretends to solve is to throw into the
dark apartments as much light as possible,—all the
light, indeed, which is visible from the window, excepting that which is necessarily lost in the process.

"If in a very narrow street or lane, we look out of a
window with the eye in the same plane as the outer face
of the wall in which the window is placed, we shall see
the whole of the sky by which the apartment can be
illuminated. If we now withdraw the eye inwards, we
shall gradually lose sight of the sky till it wholly disappears, which may take place when the eye is only *six*
or *eight* inches from its first position. In such a case
the apartment is illuminated only by the light reflected
from the opposite wall, or the sides of the stones which
form the window; because, if the glass of the window
is *six* or *eight* inches within the wall, as it generally is,
not a ray of light can fall upon it.

"If we now remove our window and substitute
another in which all the panes of glass are roughly
ground on the outside, and flush with the outer wall,
the light from the whole of the visible sky and from the

Y

remotest parts of the opposite wall will be introduced into the apartment, reflected from the innumerable faces or facets which the rough grinding of the glass has produced. The whole window will appear as if the sky were beyond it, and from every point of this luminous surface light will radiate into all parts of the room. . . . In aid of this method of distributing light, the opposite sides of the street or lane should be kept white-washed with lime, and for the same reason the ceilings and walls of the apartment should be as white as possible, and all the furniture of the lightest colours. Having seen such effects produced by imperfect means, we feel as if we had introduced our poor workman or needle-woman from a dungeon into a summer-house, where the aged can read their Bible,—where the inmates can see each other, and carry on their work in facility and comfort. By pushing out the window we have increased by a few cubic feet the quantity of air to be breathed, and we have enabled the housewife to look into dark corners where there had hitherto nestled all the elements of corruption. To these inmates the winter twilight has been shortened, the sun has risen sooner and set later, and the midnight lamp is no longer lighted when all nature is smiling with the blessed influences of day. . . .

" I cannot conclude these observations without referring to the use which may be made of them in our own city, notorious for the number of its dark and narrow lanes, and for the thousands of unlighted and unventilated dwellings which they contain. The devoted men who venture daily into these abodes of malaria and uncleanness, can alone describe to us the cimmerian darkness and the tainted atmosphere in which their pallid occupants live, and move, and have their being.

They alone can paint the harrowing scenes which disease and destitution present to them in these joyless homes. To what extent evils like these can be remedied it is a sacred duty to inquire.  To what extent they will be remedied by the large and expensive sanitary measures now contemplated, we do not venture to predict; but it is very obvious, that the upper and lower ends of the offensive lanes, which are to be intersected by the new streets, can derive little benefit from them in respect of ventilation, and none whatever in giving additional light to the houses which remain.  The only effectual mode of ventilating and lighting a dark and crowded apartment, is to strike out a large opening in the wall for the fresh admission of air, and to construct the window which is to close it so as to give the most copious entrance to the light of the sky.  A process so cheap, so easily executed, and so obviously effectual, ought to be the very first step in any measure of sanitary reform ; and it is clearly one which, if not effected by the philanthropy of the public, ought to be enjoined by Act of Parliament upon the house proprietors individually, or upon the citizens at large."

*En route* from Manchester he paid a visit which gave him great interest, and which revived many recollections of other days, when he had been intimately acquainted with Sir John Trevelyan of Wallington, and his accomplished and scientific family, while his son, Sir Walter Trevelyan, had been a friend and correspondent for many years.  The pleasure which a visit to Wallington always gave him, was on this occasion increased by the presence of some of the members of the old family circle, and by finding in the third generation fresh and intelligent scientific tastes.

My father's return to Allerly was a very happy event for him, in reviving his old love for that pretty spot, and giving him healthful occupation in the open air; he was constantly out directing the improvements of which the place, now overgrown with the trees which he himself had planted, stood much in need, and he never tired of the beautiful views which he thus opened up of the valley he loved so well. Its vicinity to his birth-place was also a great interest—a day's expedition to Jedburgh, and from thence up the Jed past the Allerly well and beyond Inchbonny, became one of his greatest pleasures, which he loved to share with his chosen friends and near relatives. Professor Fraser kindly sends me the following reminiscences :—

" 20 CHESTER STREET, EDINBURGH,
*June* 12, 1869.

"DEAR MRS. GORDON,—I fear that any incidents which I can now recall of the charming day my wife and I, with our eldest boy, spent with Sir David and Lady Brewster at Jedburgh, in June 1861, are almost too slight to be of service to you. Yet I recollect the history of those bright summer hours as containing some of the most pleasing experience in my life. Your father met us at Melrose, when we arrived there by the early train from Edinburgh, and after taking us over the Abbey we drove to Allerly to breakfast. I remember some interesting talk with him as we sauntered in the garden after breakfast,—about Locke, and Newton, and Berkeley. There was a remarkable portrait of Berkeley at Allerly, and he gave me some account of the way it came into his possession. We also spoke about some of the writings and history of my revered friend Mr. Isaac Taylor, whom I was then on my way to visit at

his beautiful cottage at Stanford Rivers. In the forenoon we all drove to Jedburgh, by St. Boswell's and
Ancrum Moor, when, with eager interest, he recalled
the local history and literary associations of the places
we passed. Then, perambulating Jedburgh, we explored the Abbey and the Castle. He took us to see
the house in which he was born. I recollect that
he pointed to a pane in an upper window of the house,
optical phenomena in which, observed in boyhood,
had, he said, set agoing the train of researches with
which his name is now associated. Later in the day
we drove up the Jed to a spot in a wood two or three
miles above the town, where, after some rambles, we
dined. All this is vivid in my recollection, as well as
the juvenile enthusiasm with which he threw himself into the life of that day, and also brought us to
live with him in the past of his own early years. I
have never seen the sanguine vivacity of youth better
blended with the beautiful wisdom and matured experience of age. The incidents of his early life which he
recalled were perhaps impressive from their number
and minuteness, rather than of a kind to admit of selec-
tion here. The power of local association and interest
was at any rate strongly marked.

" In August of the following year we had, I remember,
a delightful excursion with your father to Dryburgh
and the grave of Scott. My family were, as usual,
passing the autumn in Yarrow. My wife and I, with
Lord Amberley, who was then visiting us there, went to
enjoy a day at Allerly. In the afternoon we all drove
to Dryburgh. Much of the conversation was naturally
about Sir Walter. I remember your father told us that
he was in the habit of dining twice a week at Abbots-

ford, when its gay scenes and brilliant society were at
their best,—Allerly being at that time his home.    He
gave us some interesting anecdotes of Scott and Lockhart.
I know that we all returned to Yarrow, towards midnight, charmed by the intercourse and scenes of the day.

"It was in the following month, I think, that he and
Lady Brewster spent a day with us at Yarrow.    In the
afternoon we took them to Tibbie Shiels', where we had
tea, and where Sir David, in cordially greeting our venerable hostess, said that his last visit to the Cottage was in
the autumn of 1818, forty-five years before, in company
with the late Lord Napier.    She distinctly remembered
the visit, and reminded him of some incidental circumstances connected with it.    We had talk about
Christopher North, the Ettrick Shepherd, Scott, and
Wordsworth, who have made classical the little Cottage,
and lone St. Mary's Lake, with green and silent Yarrow.    I remember too that your father happened that day
to be much interested in a remarkable trial for murder,
in Sandyford Place, Glasgow, of which the newspapers
were full.    As I was not less interested, we had a long
discussion about the bearing of some of the evidence, in
which I was much struck by his logical ingenuity. . . .

"When in February last year I saw his body laid
beside the old Abbey, and beneath the shadow of the
Eildon hills, in the picturesque valley where his early
manhood and old age were spent, I felt the separation
as a personal sorrow, and mourned not less, that I should
see his face and be cheered by his cordial friendship no
more, than for the loss to the University of so illustrious
a representative, and to Christian faith and philanthropy
of one so humble and true. . . .—I am, dear Mrs.
Gordon, sincerely yours,            A. C. FRASER."

Daily drives in an open carriage were most beneficial to my father's health, and to be his companion in these Roxburghshire excursions was indeed a privilege of no ordinary interest. The vivid freshness of his memory and his love of the legendary history of the vale of the Tweed, which has just been mentioned, was displayed in every drive, and not called out only by an extra occasion. He used to tell with peculiar interest that " Sorrowless Field" was so named because in 1513 it was the only valley in Scotland where there was " no sorrow"—its every inhabitant, young and old, matron, maid, and infant going forth to the death at Flodden. He delighted in pointing out the short solitary grave on the narrow tongue of uncultivated land stretching into the meadows where was fought the battle of Ancrum Moor in 1545, still called Lilliard's Edge, in commemoration of " fair maiden Lilliard," who fought beside her lover against the English invaders, and earned the dubious fame of being a feminine Withrington. The " Eildon Tree Stone," where tradition declares that " true Thomas " of Ercildoune met the Queen of Faery Land,—the " Rhymer's Glen,"—the Field of Flodden seen in the blue distances,—the beauties of Bemersyde,[1] with its ancient and well-known prophetical rhyme,—the Cowdenknowes with its " bonnie, bonnie broom,"—the beautiful scenery of the Yair,—the towers of Smailholm and Darnick,—Ashiestiel, Chiefswood, and Abbotsford,— and the " Fairy Dean," with its three ruined " peels,"— are but a few of the localities which gave him the vivid pleasure and interest which he never failed to reproduce

---

[1] " Betide, betide, whate'er betide,
    Haig shall be Haig o' Bemersyde."
                                    *Thomas the Rhymer.*

in others. He read a paper at the Royal Society of Edinburgh in 1865, upon small but very interesting mineralogical formations found in the last-named beautiful valley, of which he gave the following interesting description:—

" On the banks of the Elwand Water, which runs into the Tweed, about two miles above Melrose, there is a picturesque glen called the Fairy Dean, which has become a favourite place of resort, from its association with the incidents in *The Monastery* by Sir Walter Scott. It has acquired an interest of a different kind from certain mineral concretions which have received the name of *Fairy Stones*, from their being found in that part of the rivulet which runs through the Fairy Dean.

" When the Waverley Novels were not acknowledged by their author, facts or incidents to which they referred, were always welcome subjects of conversation at Abbotsford ; and on one occasion when I happened to mention that singular stones were found in the Fairy Dean, Sir Walter Scott expressed a desire to see them, and to know how they were formed. I accordingly sent some young persons to search for them in the bed of the rivulet, and I was fortunate in thus obtaining several specimens of great variety, and singular shape, and showing, very clearly, the manner in which they were formed.

" It did not then occur to me that a description of these stones would excite any other than a local interest; but, some years ago, when in company with our distinguished countryman Mr. Robert Brown, the *Botanicorum facile princeps* of Humboldt, he asked me to accompany him to his museum, to see some remarkable mineral productions which had been sent to him, and which he had not seen before. These minerals were exactly the same as the Fairy Stones from Roxburgh-shire, but none of them were so remarkable, either in their shape or their mode of formation, as those which I now present to the Society.

" It is obvious, from the inspection of the specimens on the table, that the fairy stones are formed by the dropping of water containing the matter of which they are composed. . . . According to a rough analysis, which Dr. Dalzell has been so good as to make for me, the specific gravity of the fairy stones is 2·65, and their odour, when breathed upon, argillaceous. They effer-vesce with mineral acids, and contain the following ingredients proportionally in the order in which they are written :—*alumina, silica, lime, magnesia, oxide of iron*, and a trace of *manganese*. The black coating on many of these stones, which is too minute for analysis, and which may be easily removed, is very remarkable. If it is not carbonaceous it must be an aluminous depo-sit, when the particles of the aluminous solution have become so small as to be unable to reflect light."

In 1861 my father's portrait was introduced into a large and popular picture, under circumstances which he thought it necessary to take notice of; and he accord-ingly sent the following letter to the artist :—

" ALLERLY, MELROSE, *June* 27, 1861.

" SIR,—I have only this moment seen, in the *Times* of Monday, an advertisement of your picture entitled the *Intellect and Valour of England*, in which I am represented as announcing the discovery of the Stereo-scope. I think it right to state to you that I am not

# 346 HOME LIFE OF

the discoverer of the Stereoscope. I am only the inventor of the Lenticular Stereoscope now in universal use.—I am, Sir, yours most truly,

DAVID BREWSTER."

"To THOMAS JONES BARKER, Esq."

A copy of this letter my father carried about with him constantly in a small brown purse, from whence he took it to show his daughter-in-law, though without mentioning his reason for thus carefully preserving it. The history of the Stereoscope is curious, and was given to the public in a popular treatise on the subject, which he published in 1856. The word itself is derived from two Greek words, $\sigma\tau\epsilon\rho\epsilon\grave{o}s$, *solid*, $\sigma\kappa o\pi\epsilon\hat{\iota}\nu$, *to see ;* and the instrument " represents in apparent relief and solidity all natural objects, and all groups or combinations of objects, by uniting into one image two plane representations of these objects or groups as seen by each eye separately." Binocular vision, or the fact that " the pictures of bodies seen by both eyes are formed by the union of two dissimilar pictures formed by each," is no modern discovery. Euclid, the mathematician, Galen, the celebrated physician, Baptista Porta, a Neapolitan writer upon optics in 1593, Leonardo da Vinci, Francis Aguilonius, a learned Jesuit, who wrote in 1613, and several other more modern writers, have, either by inference or direct statement, showed their knowledge of this important optical law. Mr. Elliot, a teacher of mathematics in Edinburgh, had turned his attention to the subject in 1823, and in 1834 he projected an instrument for uniting two dissimilar pictures. He did not, however, actually construct it till 1839, when he exhibited it in Liverpool. This was a very simple in-

strument, without mirrors or lenses, and the eyes alone
being the agents, it is called the ocular stereoscope;
two pictures taken on the binocular principle were
placed at the end of a box, and represented a single
landscape in relief.    Mr. Elliot was unaware at the time,
and for years after he exhibited his discovery, that Professor Wheatstone in 1838 exhibited a stereoscope of
far higher construction,—the first Reflecting Stereoscope,
·—an instrument fitted up with mirrors, by which
the binocular pictures were made to coincide.    I well
remember the arrival of one of these somewhat cum-
brous but very interesting instruments, and the delight with which my father watched and studied its
wonderful cubes, steps, and pyramids.    Having found
in the formation and size of this instrument various
defects which unfitted it for general use, the idea struck
him of uniting the dissimilar pictures by lenses;
and it was speedily put into practice.    Mr. Loudon,
an optician in Dundee, executed several "lenticular
stereoscopes," while the application of photography to
its beautiful uses was the next and easy step.    This
form of stereoscope was exhibited by its inventor to the
British Association in 1849, but was not then taken up
in England.    In 1850 my father took to Paris one of
the Dundee stereoscopes, with photographic portraits,
landscapes, sculptures, and buildings.    M. Soleil and
M. Duboscq, the eminent Parisian opticians, saw at once
the value of this popular adaptation, and their beautiful
lenticular stereoscopes, along with binocular photographic slides of all descriptions, caused a great sensation,
and crowds flocked to see them.    It was not till the
Exhibition of 1851, when M. Duboscq sent one, amongst
other philosophical instruments, that they became

known in England.    The French stereoscope attracted
the attention of the  Queen, and M. Duboscq manufac-
tured one, which Sir David Brewster presented to Her
Majesty in the name of the maker.    On the other
hand, the reflecting stereoscope has been so thoroughly
supplanted by the other more attractive and portable
form, which has thus " *vulgarisé* "—to use the untrans-
lateable French term—the instrument all over the world,
and has indeed been so little seen, except by the
scientific, that to speak of my father as the inventor of
*the* stereoscope became a very common error.    When
any of his friends or family fell into it, he used to correct us, saying, " You mean the inventor of *this* form,"
or " of the lenticular stereoscope."

In connexion with this subject I may mention casu-
ally the rather celebrated Chimenti controversy, which
excited much discussion in the photographic world.
Dr. John Brown and his brother, Dr. Crum Brown,
noticed in 1859, in a museum at Lille, two curious drawings of a man sitting on a low stool, with a compass in
one hand and a string in the other.    They appeared to be
similar, and were placed together as if intended for the
stereoscope.    They were designed by Jacopo Chimenti da
Empoli, a painter of the Florentine school, who was born
in 1554, and who died in 1640, thus living and working
within the period that Baptista Porta's optical writings
were known in Italy.    Dr. Brown thought, on closer
inspection, that he could discover certain slight differ-
ences, indicating that the pictures must have been taken
from slightly different positions of the eyes, and even
that he could succeed in uniting the two so as to produce a decided stereoscopic effect, an experiment which
he communicated to Sir David Brewster.    With great

difficulty photographs of these drawings were obtained.[1]
The usual process of a controversy went on; some saw
no difference in the two pictures—others saw very
minute difference, which might be accidentally produced
—although they could not solve the mystery of two
commonplace drawings so nearly alike being thus
placed side by side. Others again saw discrepancies
so decided as could only be rationally accounted for
by scientific intention. Sir David held the latter view
strongly; and Professor Tait, in the volume referred to
below, gives his own opinion as follows :—

" I have very carefully considered the Chimenti
sketches, and I have concluded that they *must* have
been drawn *with the intention* of making pictures of
the same object from SLIGHTLY different points of view,
in other words, for a stereoscopic effect. From this it
would appear impossible to think otherwise than that
some form of the stereoscope, or (as is more probable)
some equivalent form of squinting, was known to the
artist. Several competent authorities, whom I have
consulted, entirely agree with me on this point."

In 1862 the Inventors' Institute chose Sir David
Brewster as their first President. Himself an inventor
of no ordinary success in the higher sense, and of no
ordinary want of success in the lower and commercial
meaning of the word, he threw himself with ardour
into all the purposes and efforts of this noble Institute,
and sent frequent literary contributions to its organ,
*The Scientific Review.* An article on " Scientific Education in our schools," which appeared in it after his
death, was sent to the editor but a few days before,

---

[1] Copies of which are in the bound volume of the *Photographic Journal*
for 1862-64.

with these words, " I am glad to see that the *Review* is prospering. I wish I could do something to help you, but I am very unwell, and not able to write."

In 1863 Sir David was again in London, in the midst of an amount of business and occupation of all kinds which it is astonishing to recall. He came to meet us at St. Leonard's, to greet us after a lengthened residence abroad, and enjoyed much the beautiful scenery around Hastings. When with him in town afterwards, we were amazed and alarmed by the long distances he walked, the constant fatigue, and his long periods of abstinence from proper food.

He was exceedingly interested in Professor Pepper's beautiful scientific experiments of the " Ghost," and its various manifestations, and he had much interesting conversation and correspondence with the Professor, who had a series of entertainments, called " Half-hours with Sir David Brewster;" the whole thing being in accordance with the " natural magic " which he ever delighted to observe. He had himself been instrumental in raising and naming a spectre, having been the first to observe how, by a simple photographic process, a good representation of an aërial figure might be effected, which was from thenceforward sold in the shops in the form of a *carte de visite*, as " Sir David Brewster's Ghost ! "

In the spring of 1864, while residing in Rutland Street, Edinburgh, my father was attacked with one of the now too frequent seizures of prostrating illness of an apparently indefinite character ; his mind was on this occasion in a peculiarly bright and placid state, and he spoke much of the happiness which his faith in the work of His Saviour gave him when feeling so near death. On one occasion I was afraid he was over fatigued, and begged him not to speak for a little ; he complied, but added, " It delights me to speak of these things." In a neighbouring house, during his own severe illness, lay, apparently also drawing nigh unto death, his beloved daughter-in-law, Mrs. James Brewster, and his sympathy and affections were much drawn out to her, many messages passing between the two apparent deathbeds. My father, however, as was his wont, rallied completely, shaking off at least for the time all appearance of illness.

In the University of Edinburgh there was always a full share of interesting and exciting topics, in which my father ever took a most active part. He was especially interested in a professorial contest of a peculiarly animated nature, which took place a year later, and the result was very gratifying to him, being the election of Professor Oakley to the Chair of Music. Prince Alfred's prolonged visit to Edinburgh in 1863-64, was an object of interest to Sir David, who was much interested in him, partly for the sake of his father, whose memory was dear to all men of science, and partly from the intelligence which he perceived in the young prince.

The following extracts are from letters written during different visits to London :—

"ATHENÆUM, *April* 29, 1863.

" MY DEAREST JEANIE,—Dr. Lyon Playfair and I waited in the reading-room till Mr. Gladstone came to take us to the *levée* in his carriage. There are to be no fewer than 183 presentations to-day, and as Mr. Gladstone had to attend a meeting of the Cabinet, he wrote to General Knollys to ask him to make our reception early. He made it the first, and gave us the *entrée* by the private door by which the Royal party entered. After leaving the Athenæum, we encountered the row of carriages which were drawn up in Pall Mall, so that parties might be admitted in the order of their arrival. We were therefore in a difficulty, but Mr. Gladstone having told the police that we were to be admitted early at a private door, they succeeded, with much difficulty, in forcing us past the innumerable obstacles by which the street was blocked up. We were shown into General Knollys's private room, where we found Lord Elcho and one or two officers, who had received the same permission that we had got. There we waited nearly a quarter of an hour. We were then summoned to the door of the reception room, where we waited nearly another quarter. Mr. Gladstone was then called in alone, I presume from the awkwardness of keeping a Cabinet Minister waiting in the passage. Some time elapsed before we were admitted. The attendants of the Prince and Princess were not numerous. Mr. Gladstone presented Dr. Playfair and me to the Prince and Princess. I placed in the Prince's hands the address, saying that it was from the Senatus of the University of Edinburgh. Mr. Gladstone, as Rector, presented the address from the students, and we retired with the usual formal obeisances, during which I found my back once or twice in the wrong direction. The Princess is truly beautiful and most intellectual-looking ; but I was told she varied very considerably, and this accounts for the different characters of her photographs."

"ATHENÆUM, *May* 1, 1863.

" I went to the Polytechnic last night to see the ghost, which is wonderful. I introduced myself to the lecturer, Professor Pepper, after the lecture, when he took me behind the scenes and showed me how the wonderful effect was produced. It is the invention of a Mr. Dircks, a civil engineer, who communicated the secret to me four years ago, and sent me a model apparatus for showing it by means of the dressed little figures I gave to Connie, and which I wish carefully preserved. I went again to the ghost to-day with Mr. Hayward, who was astonished and delighted with it. . . .

" I am to dine with Kinglake and Hayward to-day. You will see in the *North British Review* an article on Kinglake's book by Mr. Hayward, which will excite great notice."

"ATHENÆUM, *June* 5, 1862.

" I have just received from Professor Becquerel, from Paris, a copy of the Solar Spectrum, taken photographically upon plated copper. It shows all the colours, and it is now probable, being certainly possible, that we shall have in our photographs the colour of the landscape, and the tints of the human face."

"ATHENÆUM, *June* 6, 1864.

" I begin this letter at mid-day, lest I should be prevented by business from getting back here in time for the post. I am to meet Mr. Marsden Latham the Secretary, and Mr. Richardson the Vice-President, of the Inventors' Institute, on my Lighthouse affair at three o'clock, and may be detained there. After an early dinner, I walked last night to the Polytechnic, to see the improved ghost, which is charming. The scene is

z

laid at the bottom of the sea, on the stage, the upper surface of the water being represented by thin and narrow folds of the finest gauze. The ghosts of fish and mermaids swim about with great activity, and dance and frolic with a real man, who goes down among them in a diving-bell.

" Professor Pepper took me down to see all the apparatus by which the ghost scenes and dissolving views are produced, and I was introduced to the originals of the fish, mermaids, and other ghosts, that are so amusing to the public. The house was crowded, and I am told it is so every night. . . . To-morrow at twelve o'clock, as President of the Inventors' Institute, I am obliged to give evidence, before a committee of the House of Commons, on the subject of a building for the Museum of Patents,—a duty which I dislike very much."

*"June* 19, 1864, *Sunday.*

" Although I wrote to you yesterday, and have therefore nothing to tell you, I cannot help sitting down and putting myself *en rapport* with you and Connie. . . . The day has been so sultry that I went to the nearest church, St. James's Chapel, close to Jermyn Street, and heard a very nice sermon by the Rev. Mr. Oakley, on ' Ye are the light of the world,' in which he compared the properties of natural with spiritual light."

" ATHENÆUM, *June* 20, 1864.

" I observe in to-day's *Times* that Captain Palisser's invention of chilling cannon balls by melting them or rather pouring the cast-iron into cold iron moulds, has been most successful ; and that his other invention of utilizing old guns by lining them with rifled tubes, is highly appreciated by the military authorities. This news will delight the Fairholmes. This fine day, and the sight of green trees from the Club windows, make me pine for Allerly, and our visits to Parkhill and Belleville."

Shortly after the last date, a neglected cold fell heavily upon the aged frame; his state rapidly became so alarming that Lady Brewster was telegraphed for, and he was for some time under the kind medical superintendence of Dr. Sieveking, who pronounced that there was, and must have been for many years, organic disease of the heart, causing most probably much of the prostration of strength from which he had so often suffered. After some time of the most tender and careful nursing, he again recruited sufficiently not only to return to his own home, but to undertake the same summer the further journeys northward to visit his son and daughter in their Inverness-shire and Aberdeen-shire homes, which he made with his wife and little girl every year, and which were a source of much health and enjoyment to him.

The recollection of his genial happiness in his visits to Parkhill is so vivid that I cannot but allude to it. And as I shall not return to the subject, I may mention that, in looking back, it is difficult to separate one yearly visit from another, so invariable was his cheerfulness and enjoyment of the northern summer and scenery.

" Coleridge says, that every man should include all his former selves in his present; as a tree has its former year's growth inside its last; so Dr. Chalmers bore along with him his childhood, his youth, his early and full manhood into his mature old age. This gave himself, we doubt not, infinite delight—multiplied his joys, strengthened and matured his whole nature, and kept his heart young and tender ; it enabled him to sympathize, to have a fellow-feeling with all, of whatever age."[1] These observations were never better exemplified than in Sir David Brewster, who was still a child in heart ; his enjoyment of every excursion, every pleasant acquaintance, and his active share in every social amusement were as childlike as ever. On the croquet ground he showed the early determination to excel which so characterized his youth. It was a mode of taking air and exercise which particularly suited the feeble state of his limbs, which now prevented him from going to any distance without the possibility of frequent rests; and he accordingly gave himself to the game with a wholehearted energy which many will recollect ; his mortification at a bad hit, his reasons for the failure, and his determination "to do better next time," were identical with the adventures of the volunteer-ground and the Highland moors, while the strength of his expressions called forth many a smile, as he declared with all seriousness that " to croquet a neighbour's ball was a most immoral action." The first season that he was unable to play his quiet game at croquet, which was not till 1867, gave a pang to his watchers that those who despise the game could scarcely understand.

During his later visits to Parkhill his attention, like that of Newton of old, was much engrossed by the examination of the soap-bubble—an employment which was a constant delight and surprise to children and young people, who evidently thought that the philosopher was "playing himself" as well as amusing them, while the

[1] Dr. John Brown in *Horæ Subsecivæ*.

demands on housemaid and storeroom for various kinds of soap excited surprise in other regions. The table at which he sat, with his little wire cups, cubes, and variously-shaped vessels, whence he suspended or sent up the lovely filmy forms, was always a centre of attraction, while his clear quiet explanations of the beautiful phenomena were intelligible to all.

A great love of pictures of all kinds was a strong element in my father's mind—he purchased them far beyond his means, and surrounded himself with them as with familiar friends ; this was, of course, combined with an intense delight in the arrangement of them, and many were the changes of position which took place in his own large collection,—a better light, a new harmony of colour, being an ever new excuse for arranging his favourites anew. A collection of family portraits at Parkhill having been taken down to be revarnished, presented an irresistible opportunity for gratifying this taste ; the re-hanging of these pictures was therefore confided entirely to him, and his happiness in doing so, and the constancy of the low whistle of satisfaction were pleasant to see and hear. The best lights were chosen, and the proper harmonies of size and frame were carefully attended to, without due investigation of relationships. When it was afterwards pointed out that the wrong husbands and wives were occasionally grouped, his son-in-law refused to have any change made, so much did he value the handiwork of those pleasant " hanging-days," which could return no more.

" A day at Banchory" was one of the pleasant elements of all his Aberdeenshire visits. There was much that was congenial, and indeed very similar, between the characters of David Brewster and Alexander Thomson of Banchory, and a warm mutual affection existed between them. There were the same scientific tastes, the same genial humility, the same versatile knowledge, the same power of communicating it, latterly the same political sentiments, and above all the same Christian hope. The hours of the bright summer day spent in library and museum over book, microscope, shell, and prism were all too short, the only shadow being the fear of exciting and overwearying the friend who, though so much younger than my father, was in far more fragile health :—the elder one entered into his rest but three months before the other.

# CHAPTER XX.

## NOTES OF LIFE FROM 1864 TO 1867.

PHILOSOPHY consists not
In airy schemes or idle speculations ;
The rule and conduct of all social life
Is her great province.   Not in lonely cells
Obscure she lurks, but holds her heavenly light
To senators and to kings, to guide their councils,
And teach them to reform and bless mankind.
THOMSON.

FULL many a storm on this grey head has beat ;
And now, on my high station do I stand,
Like the tired watchman in his air-rock'd tower,
Who looketh for the hour of his release.
I 'm sick of worldly broils, and fain would rest
With those who war no more.
JOANNA BAILLIE.

In 1864 Sir David Brewster was appointed President of the Royal Society of Edinburgh, and on taking the chair he delivered an address, in which he gave some details of the earlier history of the Society, which were probably unknown to some of the members, who belonged to a more recent generation than their venerable President; and which included recollections of early friends and scientific correspondents, whose letters still remain for future use.   I give a few extracts :—

" In the closing years of the last and in the first decade of the present century, the Society was in a very languid condition.   In each of the years 1799, 1802, 1803, 1808, and 1809, only one of the papers read at its meetings was published in the Transactions ; while our Transactions were thus scantily supplied with papers, those actually read were few in number, and often too abstruse to excite a general interest."

It was the storm of geological controversy which was raised between the followers of Dr. Hutton and his Theory of the Earth, and the Wernerian school of Geology, which brought new life and animation into the declining Society. " The rival theories of fire and water were discussed with all the warmth, I may even say the bitterness, of political or theological controversy. Vanquished by the superior science of their opponents, the Wernerians quitted the field, and the Huttonian theory, illustrated by the eloquence of Professor Play-fair, attracted to its study the most distinguished geologists of other lands, and took a high place among the natural sciences. . . . If geology, as a science, drew its first breath within our walls, by the active labours of our colleagues, the kindred science of miner-alogy was, at the same time, earnestly studied and greatly advanced. Mr. Thomas Allan, who possessed one of the finest collections in Scotland, spared no expense in enriching it with new and rare minerals. In 1808, a Danish vessel, brought into Leith as a prize, was found to contain a small collection of minerals, which was purchased by Mr. Allan, and Colonel Imrie, a Fellow of this Society, and a contributor to its Trans-actions. Among these minerals they found a large quantity of cryolite, a substance so rare that at the market price it would have brought £5000. They found also crystals of gadolinite, sodalite, and a new mineral, to which Dr. Thomson, who analysed it, gave the name of Allanite.

" These interesting minerals had been collected in Greenland by Mr. (afterwards Sir Charles) Giesecké, during the mineralogical survey which he had made of that country between 1805 and 1813, and were shipped by him for Copenhagen in 1808. Upon his arrival at Hull in 1813, with another and more valuable collection, he learned the fate of his former specimens, and immediately proceeded to Edinburgh, where he was hospitably received by Mr. Allan, Sir George Mackenzie, and other members of this Society. During his residence here he contributed papers to our Transactions, and acquired so high a reputation as a mineralogist, that, through the interest of his friends here, he was appointed to the Chair of Mineralogy in the Royal Dublin Society.

" While the study of mineralogy was thus greatly promoted by the labours and liberality of Mr. Allan, he had the good fortune, at a later period, to bring to Edinburgh, and receive under his roof for nearly four years, á young German mineralogist of very uncommon acquirements. William Haidinger, a native of Vienna, who had studied mineralogy at Gratz, under the celebrated Frederick Mohs, came to Edinburgh in 1823, and resided with Mr. Allan till 1826, when he returned to Austria, where he prosecuted with ardour his geological and mineralogical studies, and where he now occupies a high place in the scientific institutions of Vienna. During his residence in Edinburgh he published several valuable papers in our Transactions, and delivered a course of lectures on Crystallography, at which Dr. Edward Turner and other two friends were the audience. In claiming to have been one of his pupils at these lectures, I cannot resist the gratification of claiming him as a pupil in that branch of optics, connected with mineralogy, which was then ardently

studied in every part of Europe.   When Mr. Haidinger
returned to Vienna, he prosecuted the study of physical
optics with great zeal and success, and had the good
fortune to discover one of the most beautiful facts in
that branch of science.    He was the first who observed
that curious property of the eye by which it discovers
polarised light, and even the plane of its polarisation,
without any instrument whatever.    The cause of this
remarkable phenomenon, called 'Haidinger's brushes,'
has not been discovered; but there is reason to believe
that it is produced by a structure in the retina, immediately behind the *foramen centrale.*"

Sir David either delivered or had read for him an
address each of the few years that he held his post,
and the latter part of each contained very interesting biographical notices of the members of the Society
deceased during the past year.   He frequently communicated papers, many of them on subjects connected
with the labours and discoveries of long past years.
A few of these which are before me embrace such topics
as "the Action of Uncrystallized Films upon Common
and Polarised Light;"—" On the Radiant Spectrum,"
the phenomena of which he discovered and described
to the Society in 1814;—" Observations on the Polarisation of the Atmosphere," and " Additional Observations," which were made at St. Andrews between
1841-45, where, from some of the experiments in which
he so delighted, made in "a long dark passage running
north and south," he discovered a new "point or spot
in which there is no polarisation," and to which the
French gave the name of "Brewster's neutral point;"
"Report of hourly Meteorological Register, kept at Leith
Fort, in the years 1826-27;"—" On the Pressure Cavi-

ties in Topaz, Beryl, and Diamond, and their bearing on Geological Theories ;"—" On the Structure and Optical Phenomena of Ancient Decomposed Glass"—one of the beautiful and popular subjects of investigation which many will recollect enjoying with him ; this paper is beautifully illustrated by the late Mrs. Ward, whose name has been already mentioned as a kind assistant with her most ready and intelligent pencil. Her charming drawings, correct and beautiful as they are, cannot do full justice to the strange filmy beauty which clothes this substance only in decay and death. Afterwards there is a paper upon the " Motion and Colours of Films of Alcohol, Volatile Oils, and other Fluids,"—two papers on his favourite study, the " Liquid Films of Soap-bubbles," and many others.

During my father's visit to Parkhill in 1866, an epidemic of whooping-cough was in the neighbourhood, and two cases occurred in the house ; his distress and anxiety lest his little girl should take the infection were very great, and a strict quarantine was established. It never struck any one that, while she was to escape, he himself was to be smitten down,—but so it was ; a few days after his arrival at Belleville, so severe and prostrating a cough began to tear his delicate frame, that it was quite clear that he must either have caught real or sympathetic whooping-cough, a complaint which it turned out he had never had in earlier years. He was very ill, and Lady Brewster and I were both sent for from different directions. With the knowledge of the presence of long-standing organic heart-disease, it seemed as if every paroxysm of coughing must be his last, and slender were our hopes of his rallying, while he himself was sure that the end was near. The occasional extreme

lowness of pulse, exhaustion, and abhorrence of the
necessary food and stimulants were most alarming, but
his submissive efforts to take the nourishment on which
his life depended were very touching, and were, under
God, the means of his ultimate recovery. During one
day, when he was at the worst, and occupation was
impossible, from the almost unceasing violence of the
cough, it was curious to see how his active mind employed itself in "examining" the laws of the malady,
which was so severe upon him that he would rather, as
he said, with his quaint force of expression, "have been
sentenced to ten years of penal servitude;" he counted
the seconds of intermission, calculated the approach and
duration of the fits of coughing, described clearly his
symptoms, and anxiously inquired into the sensations
of one or two other friends who in other parts of the
country had been prostrated in ripe years by apparently
the same juvenile complaint. Just before, he had received the sad intelligence of the death of his beloved
daughter-in-law, Mrs. James Brewster, who sank under
protracted sufferings, borne with the beautiful patience
of those who know and love the skilful Hand which
smites and heals.

Notwithstanding these depressing circumstances, my
father's mind was never brighter, clearer, or more active.
A favourite young scientific friend, Mr. Francis Deas,
was staying in the house at the time, and after hours
of fatigue and suffering, it was positive enjoyment to
the invalid to make the little preparations for his visit,
which was quite the event of the day. Believing himself a fast dying man, he left many instructions with
Mr. Deas as to the arrangement of his scientific instruments, etc., and two years afterwards, when the call

really came, it was to this gentleman that he confided the finishing and reading of a paper for the Royal Society which weakness prevented him from completing. It was " On the Motion, Equilibrium, and Forms of Liquid Films."[1]

Mr. Deas writes thus to Mrs. Macpherson :—

" EDINBURGH, 22*d June* 1869.

" MY DEAR FRIEND,—I most gladly comply with your request to send you what reminiscences I have of Sir David during the three weeks that I spent with him at Belleville in September 1866.

" He had been ailing for some days, and, as I well remember, he sent for me to see him in his bedroom the very night I arrived. I found him dressed, sitting at the table, busy experimenting on a subject which had recently attracted his attention, and which formed one of the principal matters of scientific interest with him from that time till his death, viz., the phenomena displayed by liquid films without sensible gravity. The subject had been shortly before brought under his notice by an account published in the Transactions of the Belgian Royal Academy of some experiments of Professor Plateau. Sir David, while repeating these experiments, had been led to question the commonly received theory of the cause of the colours of the soap-bubble. Sir Isaac Newton's calculations for the various orders of the colour bands proceeded upon the supposition that the colours were due to the interference of the rays reflected from the surfaces of the film itself. Sir David made the discovery (which, although I believe it has been

[1] This paper, with several others, was beautifully illustrated by Miss Dickenson, then residing at Friar's Hall.

questioned, has certainly not been refuted) that the
colours originated in minute particles of oily matter
floating freely upon the surface of the film. These
experiments engrossed almost his whole time while he
was at Belleville, and formed the substance of three
papers which were afterwards read before the Royal
Society of Edinburgh (*vide Transactions*, vol. xxiv.
part 3, and vol. xxv. part 1). His illness, you know,
took the form of a harassing cough, which at his
great age of course severely taxed his strength; but
his spirit seemed to defy any such restraint, and I
think there was scarcely a day that he was not at his
work. I saw him, you will remember, almost daily,
and had a great deal of talk with him, both on various
scientific subjects and those of everyday interest, and
I remember I was struck, more than I ever had been
before, by what I think was so remarkable a part of
his character, and doubtless one cause of his eminent
success as a discoverer of truth,—I mean the marvellously keen interest he showed in every possible thing.
I daresay you remember remarking the same to me long
ago, that in driving or walking with him every few
yards of the road presented either something he wished
to look at or had something to tell about.

" I was probably more struck with this peculiarity of
his during that visit than I had been before, because I
saw more of him, and tried to amuse him during his
confinement with any subject for talk I could think of.
Anyway, I well remember thinking what a sight it was
to see the grand old philosopher, who had in the truest
sense *lived* through wellnigh a century, during which
he had so diligently and so lovingly gathered in and
employed, to the glory of God and the weal of man, the

truth and the beauty which, though lying all around, it
needs the seer's eye to see, sitting there still at his
work, with all his faculties as perfect as ever, his
memory as clear, and his interest as keen.

"And touching too, was it not, that at the close of
his long life he should be admiring with the eye of the
philosopher those very beauties of the fleeting soap-
bubble which doubtless he had so long before delighted
in with a different but not less intense interest as a
child—as grieved as ever when the poor bubble broke
and scattered its glories just as they reached perfection.

"Another matter of a somewhat different kind which
greatly interested him at that time was the Newton and
Pascal correspondence. You may remember that I
remarked to you at the time that I was amazed at the
ability with which he sifted the evidence on that
subject, and struck home at the weak points in his
opponent's armour. It was more like the way an accomplished lawyer might have been expected to deal
with it than one who had never devoted himself to the
art of debating.

"Two little incidents also occur to my recollection
which I may mention, as I think they were both very
characteristic of Sir David. One was his wakening
Lady Brewster, as he afterwards told me, to look at the
changes on a gorgeous film he had been watching for
hours. His whole heart was so in the pursuit of the
knowledge of God's works that he seemed never for
a moment to doubt that any one could but go into
raptures like himself over a thing of beauty, even at
untimely seasons.

"The other incident was no less like his own kind
beautiful self. I had let a box of microscopic objects,

which had taken me months to prepare, fall down-stairs, to their utter ruin. He spoke to me so kindly and sympathizingly, saying it was like the story of Newton's papers being burnt by the upsetting of the candle, and so encouraged me that I set to work that very day to repair the loss.

"If you think this letter can be of any service to Mrs. Gordon, please send it to her; and if I can contribute anything more that may be thought of interest about one who always showed such a kindly and considerate interest in me, it will afford me the greatest pleasure to do so. I owe him at least this debt of gratitude, that I believe that, but for the kindly encouragement and assistance that—prince of philosophers as he was—he always condescended to give me in scientific subjects, much of the interest I have in those things to which he so ardently and lovingly devoted his life might have lain dormant.—Believe me, ever most truly yours, FRANCIS DEAS."

The complaint gradually subsided, and when he was able to return home it was with little apparent mischief, except an increased delicacy of the bronchial tubes.

There was one subject which loomed through half a century of Brewster's life, and which to its last months caused him so much overwhelming anxiety and distress, that it would be impossible to write any sort of faithful memoir of him without at least alluding to it. I refer to what is called "the Lighthouse Controversy." For obvious reasons my sketch of it shall be as uncontroversial as possible.[1]

[1] As I was afraid of overstating the case, I have taken the leading facts from a statement of Dr. J. H. Gladstone, Ph.D., F.R.S., which, it was well known, my father did not consider a favourable one.

There has seldom if ever lived an inventor who stood quite alone in his invention. One man has generally laid the unseen foundation, another builded thereupon, while perchance a third has put on the pyramidal head-stone which crowned it with use and glory. Nay, as the builder builds, it is not always with his own materials. The blessing upon the inventor is still due to him who in building the edifice of his invention, has the adaptive genius to utilize the stone, the lime, the lintel, and the beam of some former edifice scattered unused and uselessly around.

In 1811 Dr. Brewster had busied himself in experimenting upon a form of lens which was not at that time known in England. This was called Buffon's lens, invented by that philosopher while making experiments with the burning mirror of Archimedes. It consisted " of one piece, in which all the glass was ground away which was not necessary either for converging rays to a focus, or throwing them from a focus into a parallel beam." In 1788 Condorcet, in his " Éloge de Buffon," suggested the important improvement upon his lens of making it of several zones or circles of glass instead of one, or to use his own words :—" On pourrait même composer de plusieurs pièces ces loupes à échelons ; on y gagnerait plus de facilité dans la construction, une grande diminution de dépense, l'avantage de pouvoir leur donner plus d'étendue, et celui d'employer, suivant le bésoin, un nombre de cercles plus ou moins grand, et d'obtenir ainsi d'un même instrument différents dégrès de force." [1]

---

[1] " Éloge de Buffon," by Condorcet, quoted in *Rudimentary Treatise on the History, Construction, and Illumination of Lighthouses*, by Alan Stevenson, C.E.

2 A

Buffon's lens was intended for the purposes of heat. It was useless for the practical purposes of illumination, because "the thickness of the glass at its central part was so great as either to absorb the light by its colour, or refract it irregularly by its want of homogeneity;" and if bad for light, it was still worse for heat. Condorcet's "circles" possessed the same disadvantage, though to a lesser degree, and neither form of lens was ever executed on a large scale. In 1789 there was, however, an attempt made to apply lenses combined with mirrors for illuminating the Lower Lighthouse in the Isle of Portland, which completely failed; and why? Simply because the invention of a lens fit for practical purposes had not then been made.

In 1811 Dr. Brewster, during the course of experiments to which I have alluded, invented a lens which he called the "built-up lens." It was constructed in zones or "circles," which, it is true, was suggested by Condorcet;—although being unknown to Brewster, even this step was an independent discovery. But it did not stop there,—each zone of Brewster's lens was formed or "built up" of separate segments, and hence it received the name of the polyzonal lens,—a most important improvement for obviating the great practical difficulty in the use of lenses for illumination, which was that of obtaining pure and homogeneous glass in masses sufficiently large. In order to give increased power to his new form of lens, Dr. Brewster connected it with "an entirely new lenticular apparatus, consisting of small lenses and concave and plane reflectors, for concentrating in one point or focus the light of the sun, or for throwing into one parallel beam all the rays of light that diverged from that focus, as represented by a lamp."

Both of these inventions were intended at the time solely for the purposes of combustion. They were described and published, with clear and correct engravings, in 1812, in the article " Burning Instruments " in the *Encyclopædia*, copies of which were sent to the library of the French Institute and to M. Biot in 1815. The laws of heat and light being identical, it would have been passing strange if the easy application of the new lens and its companion apparatus to illumination, had not soon occurred to the fertile and practical mind of Brewster. That it did so, and as early as 1816, when, by his own recollections, he made the proposal of so using it to Mr. Robert Stevenson, the Engineer of the Scotch Lighthouse Board, is therefore extremely probable, but as there are no documents *directly* establishing this earlier period, it is best to waive it and the following periods of 1818-19, when, as many believe, he also pressed it upon the engineer's attention. There is no doubt, however, that in 1820 he actually proposed this application of it to Mr. Stevenson. Mr. Alan Stevenson, the son of the engineer, himself gives this decided date in a pamphlet published in 1833, in which he says, in reference to lighthouses, " that on the subject of polyzonal lenses, Doctor, now Sir David Brewster, the inventor, was consulted in 1820," while a statement in a letter from Sir David Brewster to the chairman of the Board, that he had had frequent communication on this subject with their engineer in 1820, is recorded in their minutes uncontradicted.

We are told that at this time "every lighthouse in Europe was fitted up with hammered parabolic reflectors of plated copper, or with little squares of silvered glass, combined so as to form the segment of a sphere or a

paraboloid. When a lamp was placed in the focus of
these reflectors, its light was thrown into a widely
diverging beam—so attenuated by its divergence, and
by the imperfection of the surfaces which reflected it,
that it ceased to be visible at great distances, and was
incapable of penetrating the fogs so prevalent at sea."
Very different from the "full swelling beam of light"
that might be cast upon the waters by the polyzonal
lens and its "whole light" apparatus. Mr. Stevenson,
however, did not feel convinced by the inventor's arguments, and again and again his proposals fell to the
ground.

In the meantime an important life of science was
springing up in France. On the 10th of May 1796,
Augustin Jean Fresnel was born at Broglie in Nor-
mandy. He was the son of an architect, and he was
educated as a civil engineer. In 1814, at the age of
eighteen, he is recorded to have asked his uncle the
meaning of the words "Polarisation of Light,"—the
explanation struck his vivid mind, and from thenceforward he made the rapid progress in scientific attainments which often marks the men of short career.
In 1817 he was much engaged in experiments upon
Light; in 1819 he was put upon the French Lighthouse
Committee, and turned his attention to improvements in
illumination; experiments on these were made in 1821,
and in 1822 he read before the Academy of Sciences a
memoir on "A New Method of Illuminating Lighthouses," describing as inventions of his own not only a
polyzonal lens, but, as it was afterwards called, a holophote or "whole light" apparatus of lenses and mirrors,
to be connected with it, so identically the same as those
described and engraved in the *Encyclopædia*, that it is

difficult to believe that he had not seen the article on
Burning Glasses, which had been in the library of the
Institute for six years. That he was an independent
inventor, however, is universally believed, and my father
wrote as follows of him :—

" I have always considered that distinguished philosopher as an independent inventor of the built-up lens
and its relative apparatus, as described by himself,
though I preceded him by many years; and if his
friends shall think it just to assign to him a higher
place than mine, it will be done with the delicacy and
tenderness which honourable men feel to a rival, and
with that love of truth which his colleagues and mine
in the Imperial Institute of France will not fail to
cherish and observe."

M. Fresnel sent copies of his memoir to Sir David
Brewster and Mr. Stevenson. His invention possessed
one great advantage,—it was made in a country where
the clairvoyance necessary for a foresight of success
is not only possessed in a peculiar degree, but is
always promptly acted upon. In 1822, therefore, M.
Fresnel was made secretary of the French Lighthouse
Board, and in the same year the Tour de Corduan,
at the mouth of the Garonne, was lighted by the polyzonal lens and its attendant apparatus,—while before
his death in 1827, M. Fresnel was a man whom his
nation delighted to honour as the inventor of the life-
saving instruments which guard the rocks and the
reefs of the whole French coast. It was otherwise in
England. In 1820 at the latest, the Dioptric apparatus, as the process of illumination by lenses is called,
in distinction to the Catoptric or reflector system, was
proposed for lighthouses by Sir David Brewster, being

exactly the reverse arrangement of the same invention for heat. From that time, on to 1822, when the success of the same system was shown in France,—on through fifteen years, while precious lives were being saved in France, Germany, Russia, and Holland, precious lives were being lost in Great Britain by the unaccountable dilatoriness of officials, and the dislike of the nation to "any new thing." Month after month, year after year, Sir David Brewster threw himself into the cause with the ardour and singular persistency of his nature, only to be baffled and buffeted. The history of these singular and unwearied efforts will still, it is hoped, be presented to the public, so that it is unnecessary to say more than that they were not crowned by success until 1835, when the new lighthouse of Inchkeith was illuminated by the Dioptric lights ; and their triumphant superiority so completely manifested, that since then all the new lighthouses of Scotland, England, and Ireland have been fitted up with the lenticular apparatus, although lighthouses erected previous to 1835 still send out their old feeble lights, ships struggling beyond their limited gleams still break upon the rocks, and lives still go down into the great deep. Sir David Brewster wrote :—

"If the security of life and property at sea is a subject of national concern, the expense of substituting a combination of lenses and mirrors in place of reflectors does not merit a moment's consideration. The saving of oil and other materials would go a certain length in defraying it, and the reflectors themselves might find an appropriate application to various purposes on our coasts, or on ships at sea. Who can look at the Bell-Rock Lighthouse, erected at the expense of £61,231, without lamenting that so noble a watch-tower should be still lighted by hammered reflectors, and with a distinguishing light which science and common sense have equally condemned? An expenditure of little more than £1000 would make the lighthouse a rival of the Great Corduan on the coast of France, and carry its distinctive character over a much wider range of the German Ocean. What would be said of a great railway company, if, when safer locomotives, stronger carriages, superior brakes, and surer signals have been invented and in use, they should introduce them only into their new lines of railway, and allow the passengers on the old lines to run all the risks of imperfect and antiquated machinery?"

Still it was much to accomplish, and the proved superiority of the new to the old system caused bursts of enthusiastic appreciation even from those whose prescience had been at fault for so many years. A word of gratitude from his country might therefore have been expected to the man who invented, proposed, and almost forced the adoption of the system; but the new mode of illumination was and is called the "French system," the lens impressed with the name of "Fresnel," and the Holophote with that of Stevenson. That during the years before and after the tardy adoption of the Dioptric lights, there was much of acrimonious and personal spirit cannot be denied, and Brewster's undoubted share of the blame is deeply to be regretted, while those who loved him best would fain have to remember that he had calmly submitted to want of justice rather than that he failed in the application for it. But the desire for a recognition of his services in this noble cause by his country appeared almost as if it were engraven on his heart, like the name of Calais on the queenly heart of old. His own intense dread of some day perishing in the wild waters probably made him still more tenacious of his glory in being the undoubted means in some ways at least of a saving of life, the amount of which, though it might well be greater, will not be fully known till the seas of the past and the present give up their dead.[1] After many applications, it was not till August 1867 that the Government of Great Britain finally refused all recognition of his claims. Two circumstances are pleasing to recall,—one is, that in his obituary notice in 1866, before the Royal Society of Edinburgh, of Mr. Alan Stevenson, all painful recollections were cast to the winds. His eminent qualifications as civil engineer, author, and classical scholar, are given their due place, and his death mentioned as that "which a Christian should die." The other was that, on the 17th of February 1868, the representative of the Stevenson family made before the Royal Society a few remarks prompted by good taste and feeling, in which he expressed his cordial sympathy with the recognition made of their late President's " numerous and valuable contributions to science and literature," and seconded the motion to this effect. When after his death the tardy recognition of his Lighthouse services was made in many of the newspapers, it was touching to hear the weeping remark made from sure knowledge of the hold it had upon his

[1] I cannot forbear quoting the following gratifying sentence from the eloquent Opening Address of Sir Alexander Grant, Bart., my father's successor as Principal and Vice-Chancellor of the University of Edinburgh, delivered Nov. 2, 1869 :—" Every lighthouse that burns round the shores of the British empire is a shining witness to the usefulness of Brewster's life."

heart :—" O that he could have read that, how pleased he would have been !"

The names of many eminent men, and their opinions upon this subject, all favourable to my father's claims, are in print before me ;[1] such as Sir William Snow Harris, Sir John Herschel, Sir William Thomson, Lord de Mauley, Professor Fuller, and others ; and I may be forgiven for inserting the following valuable opinions, as well as an unpublished letter from Lord Brougham to Lord Palmerston :—

" *Opinion of* DR. LYON PLAYFAIR, C.B., F.R.S., *Professor of Chemistry in the University of Edinburgh ; and* P. G. TAIT, M.A., F.R.S.E., etc., *Professor of Natural Philosophy in the University of Edinburgh.*[2]

" We have examined the written and printed documents (connected with Lighthouse Illumination) which you submitted to us, and we are of opinion that they completely justify the following statements :—

" 1. That in 1812 you suggested an important improvement on the lens of Buffon and Condorcet, such as, in fact, rendered it (for the first time) capable of being constructed. Neither Buffon's original device, nor the farther improvement of Condorcet, could possibly have been executed on a large scale.

" 2. That in the same year, you described the Dioptric Apparatus, by which the whole of a beam of light, parallel or diverging, can be thrown into a beam con-

[1] These form an appendix to Sir David Brewster's pamphlet entitled *The History of the Invention of the Dioptric Lights, and their introduction into Great Britain.*

[2] Dr. Playfair and Professor Tait's statement was drawn up in the winter of 1863-64. .

verging or parallel, constituting what has been called the Holophote.

"3. There is complete documentary evidence, that certainly in 1820, and probably even as early as 1816, you pressed on the Engineer of the Scottish Board the adoption of a Dioptric system, based on your new form of Polyzonal Lens. Fresnel's suggestion of the same improved form of lens, and of the same Holophote apparatus, did not appear till 1822.

"4. The more perfect form of Holophote is also your invention, and was not known to Fresnel.

"5. Apart altogether from the question of priority, there is no doubt whatever that it was by your persistent and unaided efforts (especially in obtaining, through Mr. Hume, a Committee of the House of Commons, after all other means had failed) that the Dioptric system was at last, in spite of long-continued opposition, tardily introduced in Britain.

"We have kept to the main facts, about which the evidence is so clear that we consider it scarcely possible for any candid mind, upon being made acquainted with it, to hesitate to give you, not only the full credit of long priority, but also the merit of having, as it were, forced this inestimable boon upon the country.

"LYON PLAYFAIR.

"P. GUTHRIE TAIT."

"*To* SIR DAVID BREWSTER, K.H."

" At a Meeting of the Council of the Inventors' Institute, held on the 30th day of June 1864, PETER WILLIAM BARLOW, Esq., F.R.S., in the chair : After hearing a Report from a Committee appointed to investigate the evidence upon the subject, it was agreed—

" 1. That in the year 1811 Sir David Brewster, adopting, as he admits, from Buffon a suggestion of a new form of lens, invented a mode of building up this lens in segments, which for the first time made it capable of construction.

" 2. That in the years 1811 and 1812 Sir D. Brewster invented and described an apparatus for burning purposes, consisting of a combination of this built-up lens with a spherical reflector behind, and small lenses and plain mirrors distributed around the focus, by which the whole of a beam of light, parallel or diverging, can be thrown into a beam converging or parallel.

" 3. That all men of science must have been fully aware, even at that time, of the identity of its effects when applied to rays of light instead of rays of heat.

" 4. That this combination, afterwards called a Holophote, embodied the fundamental features and principles of the most perfect optical apparatus for lighthouses now in use.

" 5. That the application of this combination, as peculiarly adapted for lighthouse purposes, was suggested by Sir D. Brewster to the Engineer of the Scottish Lighthouse Board as early as the year 1820, and probably much before that time.

" 6. That in the year 1822 M. Fresnel applied the built-up lens of Sir D. Brewster in combination with plain mirrors to the construction of an apparatus for the illumination of one of the French lighthouses.

" 7. That this apparatus of M. Fresnel did not involve the principle of the Holophote, nor was its application to it suggested by him.

" 8. That from the year 1820, till its final adoption and completion in a Scottish lighthouse in 1836, Sir D.

Brewster, by his personal exertions and by his writings, laboured strenuously to obtain the introduction of the Holophotal system into British lighthouses, and that it was owing to his persevering efforts in the face of much opposition that its introduction was finally effected.

| | |
|---|---|
| PETER W. BARLOW, F.R.S., | H. B. FARINGTON. |
| *Chairman.* | R. MARSDEN LATHAM. |
| JAMES GLAISHER, F.R.S. | GERARD B. FINCH, M.A., |
| J. H. SELWYN, Captn. R.N. | *Hon. Secretary.* |
| HUME WILLIAMS. | ROBT. RICHARDSON, C.E. |
| J. S. LILLIE, Kt. C.B. | (LORD) RICHARD GROSVENOR, |
| BENJ. BURLEIGH, C.E. | M.P." |
| CORNELIUS VARLEY. | |

" I have not the books in which your invention was published, but I remember distinctly reading the article in the *Edinburgh Encyclopædia* when it appeared (1812), and appreciating at once its importance for lighthouses. I felt this the more strongly, because at that time there stood in Howth a lighthouse (now demolished) provided with solid lenses, which were so thick at the centre that they did not transmit one half of the Incident light. This was a case that seemed to challenge the trial of your invention, for it would at once have removed that defect. That such trial was not made will surprise none who are familiar with the history of invention.—Believe me, etc.,          T. R. ROBINSON."

" OBSERVATORY, ARMAGH, *Feby.* 2, 1865."

" 2 *August* 1859.

" MY DEAR PALMERSTON,—As I expect to leave London to-morrow early, I have no prospect of seeing you before I go. But there is a subject on which we had so little time on Saturday, that I omitted to mention

it, though it is really one of importance and of interest
to scientific men and to the public at large.

" I mean Sir David Brewster's great services, and the
value to the country of his useful invention of Dioptric
Lights, by which an incalculable number of lives have
been saved in the prevention of shipwrecks. My
opinion, very decidedly, is that the country which most
properly showed its gratitude, and its sense of valuable
services in the case of Sir William Snow Harris's conducting rods, would be most unjust were it to pass over
Sir D. Brewster's.

" I need not remind you of the great scientific eminence of Sir D. B. He is, in fact, at the [head] of our men
of science for the originality of his researches, and the
importance of his discoveries. But the matter now
under consideration refers to a happy and most successful invention in practice, though founded on scientific
principles.—Believe me ever most truly yours,
                                        " H. BROUGHAM."


In the meantime the strength of life was waning,
though with a merciful and gentle decrease. In the
spring of 1867, my father for the first time lost his consciousness in a fainting fit which came upon him when
in a class-room of the University in Edinburgh ; assistance was at hand, and he soon rallied, so much so that,
when he came to see me half-an-hour after, he walked
up stairs as firmly and conversed as cheerfully as if
nothing had happened. When with him at Belleville
in September there seemed little change ; and when
he left it to go to the British Association at Dundee,
it was with all the want of precaution of a young
and strong man. No entreaty would make him attend

to the taking of food or rest, which might have helped him to undergo the journey more easily, and to have been more fit for the fatigues which awaited him. The heated and crowded assembly of 2000 persons, held in the Kinnaird Hall, was too much for him, almost immediately after the hurried journey. He fainted away again, and was longer in recovering than on the preceding occasion. Again he rallied, however, and was soon able to reassure Lady Brewster by writing a playful account of his horror when he became conscious of cold water, and hands kindly divesting him of his neck-cloth. After a day or two of rest, under the kind care of Mr. and Mrs. Edward Baxter of Hazel Hall, whose guest he was, he recovered sufficiently to be present as usual at his own Section A (Mathematics and Physical Sciences), where he read papers on the Colours of Soap-bubbles, on the Radiant Spectrum, on Enamel Photography, and on the Alleged Correspondence between Pascal and Newton.

Professor Balfour, the eminent botanist, one of the Secretaries of the Royal Society of Edinburgh, and the son of an early friend of my father, gives me the following notes :—" We were glad to have Sir David Brewster at the Dundee meeting of the British Association, as a noble advocate of Bible truth in opposition to the scepticism of the men of science of the present day. To see a philosopher like him, of world-wide reputation, vindicating the inspiration of God's word, and humbly receiving the truth in the love of it, was most encouraging, though we did not know that it was to be his last appearance amongst us,—the meeting at Dundee having laid the foundation of his last illness. I was on the platform when Sir David came into the crowded hall, after a most fatiguing journey, and sat down beside me. We shook hands and spoke as usual, but on looking at him shortly after I saw a change on his countenance and a shaking of the frame, which indicated faintness. He fell into my arms, and was gently laid down on the platform, means being used to revive him; he was then carried to a back room, where many of his Edinburgh friends (medical and others) rallied round him. He recovered gradually, and was glad to see so many known faces around him. When he had sufficiently recruited, he wished to go back to the room to hear the Duke of Buccleuch's address (as President), but his medical friends prevented him. After a time I got him conveyed in a cab to my friend Dr. Gibson's house, where I remained till I got him comfortably settled for the night. When sitting by him he pressed my hand, and expressed his thankfulness for the mercies he had experienced, and for the kind attentions paid to him. I committed him to Dr. Gibson, and, when departing for the night, I could not but remark the tender and affectionate manner in which he took leave of me. He thoughtfully sent me money to pay for the cab, which he refers to in the following brief but characteristic note :—

"'ALLERLY, *Sept.* 14, 1867.

"'DEAR PROFESSOR BALFOUR,—I enclose some stamps in return for a small part of my debt to you; but I know of no stamp or coin by which I can repay the great attention and kindness I received from you at Dundee.—I am, ever most truly yours,

'D. BREWSTER.'

"After the Dundee meeting, I spent a day at Rossie

Priory (Lord Kinnaird's) with some of the scientific men. Sir Charles Wheatstone was there, and told me that Sir David Brewster had gone up to him frankly at Dundee, saying, ' We have had much disagreeable discussion together, but I hope it is all forgotten now;' upon which they shook hands cordially. In repeating the anecdote to me, Sir Charles said, 'Do you really think he was sincere ?' and I could not resist the prompt answer, ' You may trust Sir David for that.'

" Sir David afterwards proposed to the Council of the Royal Society of Edinburgh that Sir Charles should be elected an Honorary Fellow. His conduct in this instance contrasted well with his reluctance some years before with regard to another honour which was proposed (though not pressed) for Wheatstone, and showed a softening of the heart which was very marked, and which could only be traced to the power of the Spirit. —Believe me, yours most sincerely,

J. H. BALFOUR."

" INVERLEITH ROW, *October* 27."

Although apparently well again, he was glad to give up other intended visits after the Association, and to hasten to his own home, which he never again left. It was not, however, to entire rest and quiet that he at first returned. One other public conflict begun before the Dundee meeting yet awaited him. I refer to the part he took about the forged correspondence between Pascal and Newton, which, if true, would have tarnished the fame of his noble and beloved master, and broken the shrine of seventy years' fervent admiration.

The discovery and establishment of the great law of universal gravitation, which was the foundation of the celebrated *Principia*, were partly, it is true, suggested by other minds, especially by Hooke and Cassini, which Newton always frankly acknowledged; but these letters, if true, would have branded the noble-minded philosopher with having fraudulently concealed that these suggestions and others had been made to him long before by Pascal. " Newton(g) dépossédé " was the heading of the first announcement of this curious "fact," which, if a " stubborn " one, would have been highly gratifying to the French nation, one of whose scientific representatives[1] had frankly confessed " that no Frenchman can reflect without an aching heart on the small participation of his own country in the remarkable achievement of the discovery of universal gravitation."

The most noteworthy part of these documents consisted in a French correspondence between Pascal and Newton, when the latter was a boy of twelve years old, in which the discovery of gravitation by Pascal is clearly announced. These were read before the Institute, and published in *Les Comptes Rendus* by M. Chasles, a French geometer and scientific historian of such undoubted character for rectitude, that even those who felt most sure of the forgery never attributed it to him. As the biographer of Newton, and the only living person who had examined his literary remains, it was natural and appropriate that Sir David Brewster should again come forward as the champion of the illustrious dead. That he did so with the zeal and the entireness of his character, and with the energy so seldom possessed at eighty-six, was only to be expected, and these qualities contrasted strongly but pleasantly with the flippancies which were opposed from unexpected quarters to his

[1] M. Arago.

intense earnestness. His acute discrimination seized at once on many points, which may be said to have proved the forgeries. Some of them may be stated thus :—

1. That there was no mention of Pascal in any correspondence of Newton.

2. So far from being a precocious boy, Newton was a juvenile idler, and when his mind did begin to work it was entirely on mechanical objects ; he himself having recorded that his first and rather clumsy scientific experiment was made when he was sixteen, namely, jumping first in one direction and then in the other in a storm, and by the measurement of both leaps, which he compared with the same leap in a calm day, he computed the force of the gale.

3. That Newton never wrote in French—a not uncommon circumstance in the history of John Bull,—but wrote in Latin to foreign savans, while it also came out that he was unable to read French without a dictionary at the age of thirty-one.

4. That Newton's mother, whose maiden name was Hannah Ayscough, and who married Mr. Smith when her son was four years old, is represented as signing her name when he was twelve years of age as " Miss Anne Ascough Newton ! "

5. That it would not have been like Newton's English mind and English pen to have professed himself " eternally grateful " even to Pascal for scientific favours, while another correspondent pointed out the amusing idea of an English boy writing of Sir Kenelm Digby as " le Chevalier D'Igby ! "

6. That the handwriting of the MS., of which M. Chasles sent a photograph to Sir David, bore not the least resemblance to the well-known autograph and

mode of signature—specimens of which were sent by Lord Portsmouth and Lady Macclesfield for the purpose of comparison.

7. That Dr. Robertson, the Scotish Historian, who flourishes in the correspondence as having been in Paris, writing in excellent French and addressing his correspondent as "mon très Honorable," has not left a single document from which it could be gathered that he was ever out of Great Britain, could write a line of French, or was in the least interested in Newton.

One name that occurred frequently in connexion with these documents was that of M. Desmaizeaux, a contemporary of Newton, and a man of whose character for probity there seem certainly to have been two opinions; upon this man, in his eagerness to demonstrate the forgery, Sir David Brewster rather immaturely fixed as the forger, overlooking the obvious inference that a man who like Desmaizeaux was "half an Englishman," and who had lived fifty-three years in England, would never have committed or allowed to pass such obvious and ludicrous Gallicisms in letters professing to be written by English letter-writers in a difficult and foreign tongue.

As these and other arguments and attacks appeared in the public papers, other and more astonishing statements were successively produced. Galileo complains of "weak eyes" in a letter, purporting to be written by his own hand, three years after his total loss of sight. A correspondence, also in French, was brought forward, said to have passed between James II. and Newton, the object of the royal letters being entirely to extol Pascal, whose scientific discoveries made him oblivious of his various heretical peccadilloes. One of these letters is

written to Newton a few days after he left his throne,
a fact announced in the meekest terms, adding that he
forgets and forgives the behaviour of Newton and the
rest of his rebellious subjects! Louis XVIII. and an
English queen or two also dash into the subject, but
all unite in the scientific glorification of Pascal.

It will never be forgotten, in connexion with this
singular episode, how many distinguished Frenchmen
came forward to rebut these accusations against Newton,
especially M. Leverrier and M. Prosper Faugère, the
latter of whom had in 1844 published in full the *Pensées*
of Pascal; he demonstrated the want of scientific probability, inasmuch as Pascal had barely been convinced
of the motion of the earth, and as one well qualified to
judge of Pascal's handwriting, he pointed out the entire
dissimilarity. Although this subject has been entirely
dropped in England, it has been again and again mooted
in France, and in its Institute, where M. Chasles still
continued his singular refusal to give up the source of
these documents, his belief in which was the only hold
they had over the French mind. The conclusion of this
story is now a matter of history. M. Chasles at last consented to disclose the name of the man from whom he
had received the Pascal-Newton papers; and, as it now
appears, of a large number of other MSS. of ancient personages of note,—amongst them Columbus, Joan of Arc,
Petrarch, Charlemagne, Julius Cæsar, etc., etc., for which
he had paid the enormous sum of £6000. The police
watched this unfortunate individual, whose chief haunt
turned out to be the Imperial Libraries, his chief occupation copying old MS., and the chief furniture in his room
chemical compounds and brown and yellow dyes! He
made confession of his guilt, and is in prison awaiting

his trial. And so ends this singular episode which caused much pain to Brewster, who to the very last was peculiarly one whose life was instinct with—

> " . . . the conscious nerve
> Within the human breast,
> That from the rash and careless hand
> Shrinks and retires distrest.
>
> The pressure rude, the touch severe,
> Will raise within the mind
> A nameless thrill, a secret tear,
> A torture undefined."

But it is pleasant to remember that he went straight from this controversy into the gathering Silences, from whose cool calm shades came the whisper, " I die at peace with all the world."

## CHAPTER XXI.

### THE END.

From earth retiring,
Heav'nward aspiring,
    All my long day's work below now done:
Calmly reclining,
All unrepining,
    Jesus, let me lean on Thy love alone.

No more low-caring,
No more wayfaring,
    These soil'd sandals loosed and flung away,
Done with the soiling,
Done with the toiling,
    All my burdens lay I down for aye.

Ended the jarring,
Past all the warring,
    Quit I gladly life's rude war-array ;
Victory crying,
Enemies flying,
    Thus my armour put I off for aye.

Earth is retreating,
Heav'n is me greeting,
    Hope is lighting up new scenes above ;
Tranquilly lying,
Peacefully dying,
    Jesus beckons upward to His love.

Head no more sinking,
Eyes no more shrinking
    From the world's gay glitter here below
Life's cup just draining,
Time's star fast waning,
    Christ Jesus, receive my soul !   To Thee I go.
                                        Bonar.

And now we come to the last days of the long working life.   My father's own expression a little later was,
that he " was an inch nearer the end every day since

Dundee," but when we were with him in October the
change was scarcely perceptible. He drove every day,
and occupied himself in showing the near beauties of
the neighbourhood to our little son and his tutor, and
arranging their more distant excursions. When we left
him it was not with more than the natural fear of what
*might* happen during the winter. To those who knew
his old fearfulness and timidity—which grace had not
up to this time fully taken away—and who saw the
great vitality and joy of work which still remained, it
seemed impossible to look forward to the inevitably near
approach of the King of Terrors without some uneasiness as to how he was to be encountered. But to those
who thus feared, it might have been said, "Why are ye
troubled? O ye of little faith!" The thoroughness of
the change that had passed upon him was yet to be
triumphantly shown, and all the fears entertained for
him were to vanish away as the mists of the morning.
We do not know much of what during the next few
weeks was passing in his mind, but his prayers were
still for the increase of faith and love, and both were
marvellously answered. A copy of *Les Adieux d'Adolphe
Monod* had been given him by the Rev. Edward Elliot,
of Brighton, whom he had great pleasure in meeting,
two years before, at Parkhill. This book he studied
incessantly, and it proved the same help to him as it
has been to many, while hearing the ever nearer
rippling of the waters of the great River. It was after
reading it that he said one day to his wife, "I feel my
faith much increased." He loved also to read and to
listen to hymns,—simple but nourishing food for philosophers as well as children. Two of his favourites
were "Rock of Ages," and "There is life for a look at

the Crucified One." Knowing that the anxious temperament still often felt burdened and wearied by temporal
cares, a well-known and simple hymn was sent to him,
of which the following are the two first verses :—

> "Oh, eyes that are weary,
>     Oh, hearts that are sore,
> Look off unto Jesus,
>     And sorrow no more.
> The light of His countenance
>     Shineth so bright,
> That on earth as in heaven,
>     There need be no night.
>
> Looking off unto Jesus,
>     Mine eyes cannot see
> The troubles and dangers
>     That throng around me ;
> They cannot be blinded
>     With sorrowful tears,
> They cannot be shadow'd
>     With unbelief-fears,"—

and this he particularly valued and often asked for.
That grace which had soothed much of the earlier
temper and some of the nervous irritability, worked also
in producing a greater consciousness of these faults.
One day he expressed to his wife his sorrow for a hasty
word, adding so touchingly, "I say these things, but I
don't mean them ; they never come from my heart."

The following letter gives so very graphic an account
of my father in these last months and days, and it
forms such a characteristic portrait of him as he was in
his unreserved moments, that I gladly introduce it
here :—

"GALASHIELS MANSE, *July* 9, 1869.

" . . . On the 22d November 1867, Sir David presided at the first meeting of the Edinburgh University
Court which I ever attended. I was struck by his

extraordinary energy, and his zeal for the prerogatives of the Senatus, which he thought some of us disposed to dispute. He turned to Dr. Alexander Wood, and subjected him to a series of searching questions as to what he would do in certain supposed cases which he held to be analogous to that about which the Court and Senatus differed; and I remember well his looking to the Lord Provost, and saying, 'When I have done with Dr. Wood, I have some questions to put to your Lordship also,' which he accordingly did. Never was he more vigorous and acute; I may add, seldom was he more excited with commonplace business; yet he not only retained perfect command of his temper, but was as courteous as possible to those of us who were against him, especially to myself, though I had opposed his views, in spite of the warm interest which he had shown in my election to the Court.

" We afterwards travelled from Edinburgh as far as Galashiels in the same carriage; and there being no one except ourselves in the compartment, he continued conversing with the utmost vigour, chiefly of the Newton controversy. I recollect saying to him that I was amazed that he did not feel fatigued, as he had come in from Allerly in the forenoon. He replied that he was wonderfully well; but that his legs and feet occasionally swelled, which indicated that his system was becoming weakened. A slight incident at the commencement of this journey was afterwards riveted in my memory. He suddenly observed that his purse was empty, and that he had nothing for the cabman, and asked me for the loan of a shilling, with which he settled his fare. I forgot all about this, and having been engrossed with parochial duty and with a business visit to London, did not see him during the following month. In the beginning of the succeeding January, I received the following (by post) :—' One shilling in postage stamps : payment of an old debt ; interest still due.— D. B.' I was astonished at his recollection of such a trifle.

" Soon after this I called on him at Allerly, and found him in a very weak state of body, but with his mental activity unabated. We had a long conversation, extending, I think, to nearly two hours. The topics he discussed were the vacancy which had occurred in the Moral Philosophy Chair (as to which he seemed determined to keep himself wholly unfettered, in the hope of getting a Professor who would maintain the honour of the University) ; the Newton affair (in illustration of which he showed me a number of curious documents) ; but chiefly the article in the lately published *Quarterly Review,* which represented Sir Walter Scott as spat upon by the Jedburgh people at the time of the Reform excitement. He protested that this was not true, though affirmed in Lockhart's *Life,* and urged me to send a contradiction to the *Quarterly.* I replied that I was quite ready to comply with his wishes, if I found it possible, but that I could not do so without further information.[1]

" My last interview with Sir David was on the 6th February 1868. Next day had been fixed for a meeting of the University Court ; and being well aware that he could not attend, I thought it courteous to call on

---

[1] Although the statement of the *Quarterly* was perfectly correct, and Sir Walter Scott was undeniably insulted at Jedburgh, yet it was not that town which was specially in fault, for the inhabitants of the whole county were assembled, and the mob from Hawick was particularly active, instigated by a Radical proprietor.

him, and to ask if he wished me to communicate anything to his colleagues. I was greatly struck by the
unfavourable change which had taken place in his
appearance within a few days. His strength was
utterly gone, and he had a death-like countenance.
Nevertheless, he was calm, clear in his intellect, and
even cheerful. He told me that he knew his time
on earth was just about to end. I said that I feared
he was right in his idea of his own condition, but
that God had been very gracious to him, in preserving his life, and his faculties also, far beyond what
was usual, and that if he was now to be removed it
must be because he had accomplished what God had
appointed as his earthly work. . . . He cordially assented
to my remarks, and declared that he was fervently
thankful to God for the blessings he had received
during so many years. Life here, he said, was not to
be wished if the mind did not retain its vigour, and he
looked for another and better life after death.

" Passing from this strain of conversation, he entered
with the utmost clearness and obvious interest upon
the business to be brought before the Court next day.
He instructed me in the order of procedure fixed by
the Act of Parliament, and specially charged me to see
that the chair was taken, in the absence of the Rector
and Principal, not by the Lord Provost, but, in terms of
the Statute, by Dr. Alexander Wood, the Chancellor's
Assessor. He then begged that I would ascertain the
general feeling about the Moral Philosophy Chair, for
which, he assured me, he continued unpledged, and
which he was very anxious might not be jobbed. ' So
strongly do I feel on this point,' said he, ' that I think
I must resign, now that I cannot attend the meetings of

Curators, that you in the Court may elect another representative in my place.' He begged me to discover whether his resignation was thought expedient, and to have no scruple about telling him the truth. I pledged myself to do as he desired, but added that I was certain we would all deprecate his giving up his Curatorship, and that there was no need for hurry, since the election would not be for several months.

" While we were speaking on this subject, he startled me with the abrupt inquiry, ' Have you ever heard any one spoken of as my successor in the Principalship?' I said, ' O no, Sir David; I am glad there is too much good feeling for that.' ' Don't tell me so,' was his answer; ' every one knows that I am a very old man, and you and I know that my end is very near; depend upon it people are speculating upon my death, just as they do upon that of an old minister, when they examine the almanac for the patron's name. I'll tell you whom I have been thinking of—Professor Christison; he is thoroughly acquainted with the College business, and would be an excellent Principal. I said so to himself some years ago; you will see him at the meeting of the Court to-morrow; just say to him what I have now said to you, and inform him that, if he wishes it, I will give him a certificate before I die.' He then enlarged on the importance of having as Principal a man wholly untainted by scepticism, deplored the prevalence of that evil among the scientific men of London, and said it would be impossible to over-estimate the mischief which the appointment to the Edinburgh Principalship of a sceptic might work. I said that certain exhibitions at the Dundee meeting of the British Association were pecu-

liarly offensive, and that I was much pleased with the rebuke administered to them by the Duke of Buccleuch.
Sir David replied that he had been delighted with the
whole of the Duke's conduct at Dundee, and particularly
with the part of it which I had specified. He then repeated the expression of his deep regret that so many
scientific men were setting themselves in opposition to
the doctrines of Scripture, and expressed his own thankfulness that he had none of their doubts about the truth
of Christianity as revealed in the Bible, but had simple
faith in the Saviour whom God had provided, and
looked to Christ alone for pardon and everlasting
life. He said that nothing but this faith could now
support him. I occasionally threw in a remark, to
prolong this deeply interesting conversation, and had
the privilege of listening to the warmest expressions of
Christian faith and hope from the aged philosopher.

" Our conversation having reverted to University
affairs, I happened to mention Professor Crawford—
' Oh !' he said, ' I know you and he are great friends ;
if you see him in Edinburgh to-morrow, I wish you
would give him my kindest regards, and say how
grateful I feel for his treatment of me. When I came
to Edinburgh, I know there was a jealousy of me in
the Theological Faculty, because I belonged to the Free
Church ; but I always met with the greatest kindness
from them, from Dr. Crawford—yes, and from Dr. Lee
also.' I answered, ' Sir David, as you have introduced
this subject, I may tell you that not from Dr. Crawford
only, but from many of the Professors in all the
faculties, I have often heard that their relations with
you were so cordial, that they despaired of ever having
another such Principal, and that they could wish you

immortal as their head.' His eyes filled with tears, and he muttered, 'I am very glad to hear all this; they and I have been very happy together, and I was most thankful to find myself Principal of Edinburgh, my own University; it was the highest office I could have aspired to, and I am now, I believe, the oldest alumnus.' 'Not quite, Sir David,' I answered; 'remember Lord Brougham.' 'Ah yes,' he said, 'he was before me, but he never took his degree, as I did, in the year—' As he spoke he rose and seemed to be moving towards a book in a shelf at the other end of the room. I interfered, begged that he would remain where he was, and offered to bring him the book. I did so, and he opened it, and stated the year of his graduation in Arts. Taking the book from him, I requested to be allowed to look into it, having some suspicion of the nature of its contents. He said, 'You will think me vain if I consent, but I cannot refuse you.' Upon examination, I found the book to consist of a collection of diplomas and other similar marks of honour which he had received. I thanked him so heartily for permitting me to see this extraordinary volume, that he said 'Now, if you are really pleased with what I have showed you, perhaps you would like to see my medals; but you will think me very vain in my old age.' I replied, 'The sight of them will be a very great favour, let me assure you.' He then rung a bell, and, on Lady Brewster answering the summons, begged her to bring his medals. She kindly complied with this request, and I had the privilege of going over, with their possessor, these tributes to his genius. We spoke of them as evidences that he had been considered useful in his generation, for which all praise was due to God. We constantly returned to the grand theme on which he was inclined to converse—that he was a dying man, and that nothing could now be of the slightest avail to him except the blood of Jesus.

" Such was my last interview with your father. It was longer than I would have ventured upon if I had not found him desirous that I should remain. I can never cease to reflect with gratitude upon the exalted privilege vouchsafed to me of witnessing what was so very nearly the close of a noble earthly career.—I have the honour to be, with sincere respect, dear Madam, yours faithfully,                              K. M. PHIN."

" Mrs. Gordon."

Professor Balfour sends me an interesting letter written about this time, with the following explanation : —" The note refers to a communication which I had made relative to some passages in Sir David's address to the Royal Society of 1867, which I thought might be misinterpreted, so as to bring out the idea that a certain preparation of the heart for the reception of Divine truth might be effected by the cultivation of science. I think he allowed me to make a slight alteration :—

" ' ALLERLY, *December* 4, 1867.

" ' DEAR PROFESSOR BALFOUR,—Many thanks for your observations on the address. In both cases my meaning is such as you would desire it to be, but I cannot bring out that meaning better than by the words I use. By the word *wise* I mean a person who simply knows that God's glory IS shown in the *heavens*. A perfectly ignorant man, however good a Christian, cannot know this.

" ' By the words *higher revelation* I of course mean the

great truths revealed in the Bible; and I assert that a young man, with scientific instruction, will be prepared by it for the reception of revealed truth, and will be better able to encounter the objections drawn from science, than if he had *not* been instructed in science. Perhaps my meaning will be brought out better by substituting for the words, " that higher revelation," the words, " those revealed truths."—I am, ever most truly yours, D. BREWSTER.' "

He still continued able to work and to move about, and was not under medical attendance—for his own doctor being absent, he refused till even nearer the end to see any one, and then only consented to have a medical man from a country town at some distance. But the conscious-ness of the last weakness being close at hand increased upon him, and the week before he was finally laid up was spent in a literal setting of his house in order, which was most characteristic of his whole past life. Lady Brewster tells me that each day of that long week was spent as if in the most active preparation for a journey. Letters were written—or dictated to his faithful com-panion, and signed by himself; papers arranged ; books put by, and after each piece of business he would say, " There THAT 'S done ;" then something else was begun and finished—not a moment wasted—no pause re-quired—not a word of what was at hand, lest either worker should break down—a strange week of patient, unwearied, accomplished work !

One of these letters was to an old and attached friend, of whose unwearied kindness and affectionate attentions he ever expressed the most grateful recog-nition. It was as follows :—

"ALLERLY, *Feb.* 2, 1868.

" MY DEAR LADY COXE,—I have for several days been proposing to write to you, but having nothing agreeable to myself to say, and nothing agreeable to you to hear, I have been silent.

" I am hardly able to walk from my library to my bed-room, and want of breath, sleep, and appetite make me a genuine invalid, quite unable to do the duties in the University were I in Edinburgh.   I regret this bitterly, as there is so much valuable work now being done in promoting the prosperity of the University.   My complaint has been advancing so rapidly as to indicate a no very distant termination, and after such a long and happy life as I have enjoyed, I do not repine that a higher will than mine should be done.   But still, though faith be strong, and the prospects of the future bright, it is difficult without emotion to part with those kind and valued friends who have performed with us the journey of life, and shared with us its joys and its sorrows.

" I need not say, my dear Lady Coxe, how much of my happiness has arisen from your kind and affectionate attention, and how sincerely I wish that your life may be as long and as full of blessings as mine has been. —With our united kind regards to Sir James and Dr. Cumming, I am, my dear Lady Coxe, ever most truly yours,                                           D. BREWSTER."

One little piece of business was the arranging that a copy of each of his works should be set apart for an " author's table" at a bazaar, the proceeds of which were to help in establishing a Medical Mission in Aberdeen.   On Friday his loving, careful wife implored

2 c

him to remain in bed; but no!—"Let me rise once more," he said; "I have still a little work to do."  On that day he dictated a farewell letter to Professor Balfour, and to the members of the Royal Society of Edinburgh.  It was as follows:—

"ALLERLY, *Febry. 8th.*

"MY DEAR PROFESSOR BALFOUR,—I have tried in vain to finish the most important of my papers on Liquid Films, but the most beautiful drawings of all the phenomena, which its purpose was to describe, have been finished.  I think therefore that my friend Mr. Deas will, by means of these drawings, produce an interesting paper.  The drawings are numerous, but many of them may be reduced by cutting off the long tails of the glass vessels, or otherwise. . . . I beg you will offer to the Council my best thanks, and accept of them to yourself, for all the kindness that I have received from you since I became President of the Society.

"I had expected to do the work of this session, but my indisposition advanced so rapidly that I found myself unfit for the smallest exertion, mental or physical.  At my great age, and with a strong faith, the change is not unwelcome.—I am, ever most truly yours,

"D. BREWSTER."

In the course of that afternoon he saw the Rev. Mr. Cousin, his own pastor, who has recorded the visit as follows:—

"The last day he was able to be in his study— three days before he died—it was my privilege to see and converse with him.  He knew that he was dying. 'My race is run,' he said; and there was something almost of the old scientific habit of thought in what he added—'From the palpable failure of strength from one day to another, I feel as if I could count the very day when all must close.' Usually he was very reserved in speaking of himself, but on this occasion his mouth was opened and his heart enlarged. He spoke with deep feeling and tenderness of the happiness he had enjoyed in life. 'Never man,' he said, 'had more cause for thankfulness than I, but with all that,' he added, 'now that I can be of no use to myself or any one else, I have no wish to linger here.' He expressed the most perfect acquiescence in the Divine will, and the most perfect peace in' reliance upon Jesus in the prospect of standing very soon in the Divine presence; 'and yet,' he added, with something like a falter in his voice, ' it is not without a wrench that one parts with all he has most loved on earth.'"

That night the work was all over, but the usual evening occupations still remained, which I cannot forbear describing as it was given to me by the third of the little group :—

" On Friday the 7th February, dearest papa's last night in his library, Connie read to him as usual, after his dinner, before going to bed, the 27th Psalm and 6th Hebrews, singing a hymn to him, as she always did, ' There is a happy land.' Previous to the reading they had two games of dominoes together. This allowance of reading, singing, and games never varied, but seeing him look tired, and knowing how poorly he was feeling, I first advised only *one* game, and then only *one* chapter, but his reply each time was, ' No, we must do all just as usual; it may be the last time.'"

The fond quiet kiss and good-night over, nothing else remained, and as he left his study he said quietly,

"Now you may turn the key, for I shall never be in that room again." When he undressed, he said, "Take away my clothes, this is the last time I shall wear them;" and when he lay down—"I shall never again rise from this bed." On Saturday, the medical attendant still thought he might rally, so wonderful was his constitution, but the dying man knew better. He was able, however, to see the Rev. James Herdman, D.D., of Melrose, who had been one of his students in the old St. Andrews days, to whom we owe the following interesting notes:—

"MELROSE, 28*th July* 1869.

"MY DEAR MRS. GORDON,—The last time I saw your father was on the Saturday evening, less than forty-eight hours before he died. I had gone over to Allerly merely to inquire after him, and was leaving the grounds when a message came that Sir David would like to see me.

"I found him in bed, very helpless, but with eye undimmed, and his face expressive of perfect peace. He spoke without difficulty, though with a little huskiness of voice. He said he was glad to see me once again, and he wished me to pray with him.

"I made a remark about his hope, when he said with emphasis that 'it was on the Rock—Christ alone.'

"Had he no doubts? no fear? 'None. The blood of the cross had washed away his sins; he had life in Christ; this he was sure of, for God had said it.'

"Had it always been thus with him? 'For long; years ago he had been enabled to trust in the Crucified One, and his confidence had never been shaken.'

"Had he no difficulty in believing all the Bible? in these days scepticism was common among scientific men. 'Common! alas, few received the truth of Jesus. But why? It was the pride of intellect—straining to be wise above what is written; it forgets its own limits, and steps out of its province. How little the wisest of mortals knew—of anything! how preposterous for worms to think of fathoming the counsels of the Almighty!'

"Sir Isaac Newton was quoted,—'I seem to have been only like a boy playing on the sea-shore, and diverting myself now and then with a smoother pebble or a prettier shell than ordinary, while the ocean of truth lay undiscovered before me!' 'Yes! yes, indeed! Yet how sad that Newton should have gone so far wrong! He was far from sound in his views of our Lord's person—in fact, they were of the Arian type. . . . The divinity of our Saviour is fundamental.'

"The wish of many to relax our creeds was referred to. 'He had no such wish; it was just an index of the restlessness of the age, and want of submission to what is revealed. He was thoroughly satisfied with our Confession of Faith.'

"Did the Christian mysteries give him no trouble? 'None. Why should they? We are surrounded by mysteries. His own being was a mystery—he could not explain the relation of his soul to his body. Everybody believed things they could not understand. The Trinity or the Atonement was a great deep : so was Eternity, so was Providence. It caused him no uneasiness that he could not fully account for them. There were secret things that belonged to GOD. He made no attempt to reconcile the sovereignty of grace with the responsibility of man; they were both true—he could *wait* to see their harmony cleared; they were not con-

trary to reason, however incomprehensible. When he
found a doctrine plainly stated in the Bible, that was
enough; God knew; he could depend on His word: we
should not expect in this world to be free from obscuri-
ties and apparent discrepancies, and things beyond our
grasp. He thanked God the way of salvation was so
simple; no laboured argument, no hard attainment was
required. To believe in the Lord Jesus Christ was to
live; he trusted in HIM, and enjoyed His peace.'

" Such is the substance of that most delightful conversation, in which was repeated grateful mention of
the Lord our Righteousness. I wrote a memorandum
of it at the time. Rarely, if ever, have I seen a more
child-like and happy faith. 'Let me die the death of
the righteous, and let my last end be like his.'—Yours
very sincerely,                JAMES C. HERDMAN."

On Saturday morning those of his family who were
within call were telegraphed for, and Colonel and Mrs.
Brewster Macpherson arrived in the evening. Owing
to the telegram being just too late for us to take the first
train from Clifton, and the scarcity of trains on Sunday,
my husband and I did not arrive till Sunday evening,
some hours later than we were expected. It was touching to find the craving of his heart for us, which he had
been expressing through the day, fearing that we should
be too late. " Oh, how I have wearied for you!" were
his simple words, and then he seemed perfectly satisfied.
His kind and much appreciated friend, Sir James Simpson, arrived with us; he found him pulseless, but the
excitement of the arrivals seemed to give him new
energy, and a perceptible pulse returned. It was, indeed, something remarkable, and never to be forgotten,

to hear the conversation between those two eminent men. Something was said of a hope that he might yet rally. "Why, Sir James, should you hope that?" he said, with much animation. "The machine has worked for above eighty years, and it is worn out. Life has been very bright to me, and now there is the brightness beyond!" Sir James' Simpson then asked if he wished any one in particular to take charge of his scientific papers; he answered, "No; I have done what every scientific man should do, viz., published almost all my observations of any value, just as they have occurred." And then came a fluent stream of well chosen words from the dying philosopher, describing a scientific phenomenon connected with one of his favourite researches, which made one breathless with astonishment to listen to. Not a mistake, not a confused word was there, except once, when Sir James gently substituted the word "white" for "black." Although already before the public,[1] the following account is so much better than mine could be that I quote it :—

"He then explained that he had left one paper on Film forms for the Society, and went on to express an earnest regret that he had not had time to write for the Society another, descriptive of the optical phenomena which he had latterly observed in his own field of vision, where there was a partial degree of increasing amaurosis, which, he thought, might be yet found a common form of failure in the eyes of men, ageing and aged like himself. He described the appearance of this partial amaurosis minutely and energetically, telling me, for

[1] Quoted from an address before the Royal Society, delivered by Sir James Simpson, Bart., February 17, 1868.

your information, that the print of the *Times* newspaper
had begun for a year or two past to look at one part in
his field of vision as if the white interstices between
the letters 'were lightly peppered over with minute
dark powder;' and this amaurotic point was, he observed,
latterly extending like the faint extending circle around
a recent ink spot on blotting-paper."

Hearing all this, and watching the play of the expressive countenance, it was almost impossible to believe
that death was or could be at hand ; and that night more
than one heart hoped against hope.   The disappointment, though felt to be unreasonable, was proportionally
great when, the next morning, before leaving Allerly,
Sir James Simpson pronounced that my father could
not live over the day.   Monday the 10th of February
was a day of suffering from weakness, breathlessness,
and that constant desire of change of position, the
varied discomforts of which so often form the principal
suffering of a deathbed.   Pain there was little of,
except occasional spasms through the chest, significant,
I suppose, of the heart disease, which, although not that
of which he died, was complicated with the pneumonia
and bronchitis, which proved the actual messengers of
death ; once faintly complaining of one of these shoots
of pain, we did not catch his words, and it was with the
energy of old that he raised his head with a glance of
amusement, spelling distinctly, "P-A-I-N."   Upon another
occasion a play upon the word he used, and a bright
cheerful smile reminded us of the old social jest and
laughter.   All fear had passed for ever.   Throughout the
day he longed for the moment of dismissal.   " When
will it come ?"—" Oh, how long it is of coming," he said
several times ; and once he said, " What hard work it is

to ' put off this mortal coil !' " For a few hours he was
very languid, but listened with intentness to every passage of Scripture repeated to him, and if he did not
catch every word he asked for it again. Once during
this time, while partially dozing, his mind seemed to
waver for a moment, and he spoke as if " Jane Maitland"—an intimate friend—was in the room, but that
was the only brief passing cloud. He sent an affectionate message to the only absent member of his
immediate family, who was far away in Burmah. His
little daughter was brought in to see him for the last
time, and repeated to him the hymn beginning with,

> " Just as I am, without one plea,
> But that Thy blood was shed for me,
> And that Thou bid'st me come to Thee,
> Oh, Lamb of God, I come."

He was very thoughtful of his loving watchers, fearing over-fatigue for them, and saying once, with such
touching sweetness, referring to this fear, and the
trouble he thought he gave, " Oh, how sorry I am for
you all !" and when assured that it was the greatest
happiness to be near him, his uneasiness ceased, and
there was but the tender pressure of the hand,—the
long earnest gaze,—the meekness with which, to please
those who loved him, he continued the difficult task of
taking nourishment. He was always peculiarly reverential and guarded in his way of speaking of Deity,
habitually using the words " God," " the Lord Jesus
Christ," " Our Saviour ;" but on his deathbed, the sense
of the nearness, and the love of the Lord Jesus, at once
his God, his Saviour, and his Righteousness, overcame
the habits of reserve of a lifetime, and he only spoke
of " Jesus" as a personal, living, waiting Friend. Once,

when a sense of difficulty seemed to cross his spirit, he said, " JESUS will take me safe through," with restored confidence. Another time, the seldom-spoken words came to my lips, and I said, " You will see *Charlie !*" and then gathering himself up after a pause, he answered, as if in gentle rebuke, " I shall see JESUS, who created all things; JESUS, who made the worlds; I shall see Him as He is;" and he repeated, with that pathetic return to his native Scotch, which was not uncommon with him when greatly interested, " I shall see JESUS, and that will be ' grand,' "[1] with an ineffably happy, cheerful look. " You will understand everything then," it was said. " Oh yes," was the answer, which seemed to come from a very fulness of content. " I wish all learned men had your simple faith," it was said at another time ; and again there was the pause and the gathering up, and the words dropped out, each with its own weight of feeling and of meaning, " Yes; I have had the Light for many years, and oh! how bright it is! I feel so SAFE, so SATISFIED."

There came a few moments when his pulse was more perceptible, there seemed a shade less of exhaustion, and it almost seemed as if he might partially rally ; but even as this whisper passed between two of the watchers the sudden change came—the fixed gaze—the rigidity of the once mobile face—the glaze over the soft blue eyes— the silver cord was loosed, the golden bowl was broken, and the spirit fled back rejoicing to Him who gave, instructed, and redeemed it. His daughter-in-law, who had never seen a soul pass away before, wrote afterwards, " I thank God I have been present when *his* passed away. The sight was a cordial from Heaven to me. I

---

[1] The meaning of this word in Scotch is much more familiar than the English word ; it is a homely expression for *delightful.*

believed before, but *now* I have seen, that Christ has truly abolished death,"—while the two that returned to a distant dwelling-place, where the daily blank and the vacant chair could not be seen, felt for long as if there had been no death, and sorrow were impossible, so much did it seem as if they had returned from the golden gates of " the City which hath foundations."

My father's abhorrence of all funeral display was so well known to his family, that, in accordance with his often expressed opinions, they intended to confine it entirely to his nearest relatives, even valued friends and connexions not being invited.   They did not, however, think it right to refuse the desire of the members of the University and Royal Society of Edinburgh to be present; while many friends and neighbours, and large numbers of the inhabitants of Melrose and Gatton-side, and some neighbouring villages, joined the procession unasked.   The shops were closed, the bells of the churches of all denominations were pealed—groups of cottagers watched the procession, some in tears, others clad in mourning, and the sympathy of the people, with whom, though he never sought popularity, he was ever a favourite, was most marked and soothing.

It was in a strange variety of the elements that he was borne to his rest from his long-loved cottage home :— between the hedgerows,—among the orchards,—past the favourite bend of the river above Gattonside, which he had never before passed with eyes dulled to its beauty,—across the bridge of Tweed,—past the Abbots-ford road,—away under the shadow of the Eildons,— into the old Cathedral burying-ground, where lay two beloved ones—on he went—while tempest, calm, snow, and sunshine, alternated through the day.   There he sleeps till the resurrection morn, and upon the stone which marks the spot where he lies, near to a sculptured window of the old Abbey, are the simple and suitable words—

"THE LORD IS MY LIGHT."—Psalm xxvii. 1.

Dr. Guthrie—my father's friend for many years—preached a funeral sermon in the Free Church of Melrose, from which I make a few extracts :—

"I thank God, with all my heart, for His departed servant, that now, when the battle is over, he fought the good fight so well, and in days of doubt, and of darkness, and of declension, kept the faith—not only kept, but clung to it. He was one who was not ashamed in the highest assemblies of the land to stand up as a Christian, and avow himself a believer. . . .

"He knew the difference, which some seem not to see, between the sphere of revealed religion and that of arts and sciences. Theirs is the region for discoveries, and new truths, and novelties—for something the world never saw, or thought of, or dreamed of before—for the progress that lies between the first log-hut which screened its tenants from the storm, and the proud palace of kings ; that lies between the path man cut in the primeval forest, and the iron road along which he skims with fire and water yoked to his chariot wheels ; that lies between those beacon fires that blazed far and near on your Border hills, carrying the news of invasion across the land, and the wires by which I flash a message from the Old World to the New, through the bowels of the mountains and depths of the sea. No doubt the progress of science, and a better acquaintance with the language and lands of the Bible, may and will throw light on some of its obscurer passages ; but men forget what Sir David Brewster knew, and what, I rejoice to say, he bore his testimony to. The arts and sciences

have new discoveries to boast of, and they will have
more discoveries to boast of, but the Church has
none. It is an old Bible; it is an old faith; it is an
old system; it is as old as the Fall, and it will last till
the end of all things. The martyrs of science, Galileo,
and such as he, suffered for new truths—the martyrs of
the Church for old ones. . . . Sir David Brewster clung
to the old faith; he walked in the old paths. A great
man beside whom your so-called philosophers, who are
assailing our old faith, are pigmies—mere pigmies; a
man whose brows not his own country only, but other
lands, delighted to crown with their highest honours;
this man of world-wide reputation—this, in some respects, the greatest of modern philosophers, observe—
and let the world observe it—was a humble Christian.
He was a devout believer; no mere speculator in
theology, but a sincere believer in the Word of God,
who came down from the very pinnacles of science to
open his Bible, and bending that venerable head over
the sacred page, he read it with all the faith of

> " ' Yon cottager, who weaves at her own door,
> Pillow and bobbins all her little store;
> Content, though mean, and cheerful if not gay,
> Shuffling her threads about the livelong day.
>
> She, for her humble sphere by nature fit,
> Has little understanding, and no wit;
> Just knows, and knows no more, her Bible true,—
> A truth the brilliant Frenchman never knew;
> And in that charter reads, with sparkling eyes,
> Her title to a treasure in the skies.'

Yes! Sir David read his Bible, and when he lay a-dying,
he said, 'I have been reading and studying the Bible
all my days, but I shall soon know more than I ever
did.' Knowing in whom he believed, he had no doubt
of that. Blessed man! Blessed faith! Blessed peace!"

By Sir David Brewster's own wish, his books, pictures, instruments, and papers, have been taken to Belleville, where a room has been devoted to the three latter. Many portraits of him remain to us—some most excellent in point of execution, and a few good in likeness, though not one represents the living, breathing, working Man. His extreme mobility of countenance and variety of expression rendered it a task of peculiar difficulty to represent him as he lives in memory. The best pictures are two by his friend Mr. Salter Herrick,—one bought by the University of St. Andrews, the other at Parkhill; two by Mr. Norman Macbeth,—one painted for the Royal Society of Edinburgh, the other at Belleville; one by Sir John Watson Gordon, which his brother Mr. Watson presented to the nation, and which is in the South Kensington Collection; and two at Allerly, by Mr. Wilson and Mr. Whiten. He was frequently made a study for photography, where the difficulty of expression was equally formidable. The photograph taken by Mr. Musgrave, which forms the frontispice to this volume, was his own favourite. A photograph by Dr. Adamson of St. Andrews, with whom he frequently worked in photography, is an admirable representation, while there are excellent ones by Mr. Rogers of St. Andrews, and a very good one, afterwards painted, by his friend the late M. Claudet. Mr. Brodie is preparing a statue of him, which is to stand in the quadrangle of the University of Edinburgh. Mrs. D. O. Hill executed some years ago a good marble bust, and there is a much valued one, also in marble, at Parkhill, given to us by him, taken many years ago, I believe, by a Fife sculptor. All these "pleasant pictures," however, with their defects and beauties, must pass away and be forgotten in the day of dissolved elements and fervent heat, but the "likeness of

God," which is now " stamped on his brow," where he sees " Jesus as He is," shall last for ever and ever.

The night he first rested in his new quiet home, the stars shone brightly out in their beauty after the stormy day ; and the whisper came from the one who knew best his unwearied habits of work, " Oh, how busy he must be !" I give in conclusion the following lines,[1] which were tenderly and solemnly penned with a loving remembrance of all the old habits of careful composition and correction so long practised and inculcated.

> Under the storm !
> Under the storm !
> Lift ye gently the aged form !
> Bear him tenderly down the stair—
> Carry him out to the wintry air !
> Let him into the shelter go
> Of the plumy pomp of the conquer'd foe !
>
> Under the calm !
> Under the calm !
> Bear him along with a victor's palm !
> Borrow a glow from the purpled dell,
> And a gleam from the river he loved so well,—
> Let the bells ring out a birthday chime
> For the soul new-born from the throes of time !
>
> Under the snow !
> Under the snow !
> Into the damps and the dews below !
> Lay him down with his long-loved dead,
> Weep ! if ye will, o'er his silver head.
> We have not an honour to reach him now—
> We have not a love that can touch his brow !
>
> Under the sun !
> Under the sun !
> Joy ! for the saved one whose race is run !
> Joy ! for the gift of the doubtless trust,
> That shall parry many a doubter's thrust.
> Joy ! for the saint with his fair white stole,
> Of Christ's finished work in the glorious goal !

[1] These verses have been beautifully set to music by Miss Helen E. B. Drummond, entitled "Under the Storm ! Under the Calm !" Published by Messrs. Paterson and Sons, Edinburgh.

Under the skies !
Under the skies !
Where the hosts of heaven in glory rise ;
They shine on the couch where the sage is laid,
To his first night's sleep in the cloister'd shade.
Doth he walk " astonied " their lands of light ?
Hath he found a hest for his spirit's might ?
Hath he lifted a beacon in space unknown ?
Hath he solved the hues of the prism'd throne ?
Hath he met with his peers of the elder days ?
Hath he learnt from the seraphs new meed for praise ?
We see not ! we see not ! but THIS we know,
He hath bowed his head with its honours low !
" Not mine ! not mine ! " is his whisper meet,
As he casts his crown at his Saviour's feet.

# APPENDIX.

In order to give some tangible proof of the extent and variety of my father's labours, I add a list of his miscellaneous writings, so far as I have been able to ascertain their nature and number :—

1. Remarks on Achromatic Eyepieces.—Nicholson, Journ. xiv., 1806, pp. 388, 389.

2. Description of a new Astrometer for finding the Rising and Setting of the Stars and Planets, and their Position in the Heavens.—Nicholson, Journ. xvi., 1807, pp. 320-324.

3. Description of a Circular Mother-of-pearl Micrometer.—Tilloch, Phil. Mag. xxix., 1807, pp. 48-52.

4. Remarks on M. Burckhardt's Contrivance for Shortening Reflecting Telescopes; with a new method of making Refracting Telescopes with a Tube only one-third of the focal length of the Object-Glass.—Tilloch, Phil. Mag. xxxiii., 1809, pp. 290-292.

5. On the Fibres used in Micrometers; with an account of a method of removing the Error arising from the Inflection of Light, by employing Hollow Fibres of Glass.—Tilloch, Phil. Mag. xxxiii., 1809, pp. 383, 384.

6. Demonstration of the Fundamental Property of the Lever (1810).—Edinb. Roy. Soc. Trans. vi., 1812, pp. 397-404; Nicholson, Journ. xxx., 1812, pp. 280-285.

7. On some Properties of Light.—Phil. Trans., 1813, pp. 101-109.

8. On the Affections of Light transmitted through Crystallized Bodies (1813).—Phil. Trans., 1814, pp. 187-218.

9. On the Polarization of Light by Oblique transmission through all Bodies, whether Crystallized or Uncrystallized.—Phil. Trans., 1814, pp. 219-230.

10. On the New Properties of Light exhibited in the Optical Phenomena of Mother-of-pearl and other Bodies to which the

Superficial Structure of that substance can be communicated.—Phil. Trans., 1814, pp. 397-418 ; Thomson, Ann. Phil. iii., 1814, pp. 190-196.

11. Results of some recent Experiments on the Properties impressed upon Light by the action of Glass raised to different Temperatures, and cooled under different circumstances.—Phil. Trans., 1814, pp. 436-439.

12. On the Optical Properties of Sulphuret of Carbon, Carbonate of Barytes, and Nitrate of Potash, with inferences respecting the Structure of Doubly Refracting Crystals (1814).—Edinb. Roy. Soc. Trans. vii., 1815, pp. 285-302.

13. On a new Species of Coloured Fringes produced by the Reflection of Light between two Plates of Glass of equal thickness.—Edinb. Roy. Soc. Trans. vii., 1815, pp. 435-444.

14. Expériences sur la Lumière.—Paris Soc. Philom. Bull, 1815, pp. 44-46.

15. Additional Observations on the Optical Properties and Structure of Heated Glass and Unannealed Glass Drops (1814).—Phil. Trans., 1815, pp. 1-8.

16. Experiments on the Depolarization of Light, as exhibited by various Mineral, Animal, and Vegetable Bodies, with a reference of the Phenomena to the General Principles of Polarization (1814).—Phil. Trans., 1815, pp. 29-53.

17. On the Effects of Simple Pressure in producing that Species of Crystallization which forms two oppositely Polarized Images, and exhibits the Complementary Colours by Polarized Light.—Phil. Trans., 1815, pp. 60-64.

18. On the Laws which regulate the Polarization of Light by Reflection from Transparent Bodies.—Phil. Trans., 1815, pp. 125-159.

19. On the Multiplication of Images, and the Colours which accompany them, in some Specimens of Calcareous Spar.—Phil. Trans., 1815, pp. 270-292.

20. On new Properties of Heat as exhibited in its Propagation along Glass Plates.—Phil. Trans., 1816, pp. 46-114.

21. On the Communication of the Structure of Doubly Refracting Crystals to Glass, Muriate of Soda, Fluor-Spar, and other substances, by Mechanical Compression and Dilatation.—Phil. Trans., 1816, pp. 156-178.

22. On the Structure of the Crystalline Lens in Fishes and Quadrupeds, as ascertained by its action on Polarized Light.—Phil. Trans., 1816, pp. 311-317.

23. On the effects produced in Astronomical and Trigonometrical Observations, etc., by the Descent of the Fluid which lubricates the Cornea.—Quart. Journ. Sci. ii., 1817, pp. 127-131.

24. On the Decomposition of Light by Simple Reflection.—Quart. Journ. Sci. ii., 1817, p. 211.

25. On the Connexion between the Primitive Forms of Crystals, and the Number of their Axes of Double Refraction (1819).—Edinb. Mem. Wern. Soc. iii., 1817-20, pp. 50-74.

26. Sur le mouvement perpétuel.—Annal. de Chimie, ix., 1818, pp. 219, 220.

27. Description du Kaleidoscope.—Bibl. Univ. viii., 1818, pp. 155-160.

28. On the action of Transparent Bodies upon the differently coloured Rays of Light (1815).—Edinb. Roy. Soc. Trans. viii., 1818, pp. 1-23.

29. Description of a New Darkening Glass for Solar Observations, which has also the property of Polarizing the whole of the transmitted Light (1815).—Edinb. Roy. Soc. Trans. viii., 1818, pp. 25-30.

30. On the Optical Properties of Muriate of Soda, Fluate of Lime, and the Diamond, as exhibited in their action upon Polarized Light (1815).—Edinb. Roy. Soc. Trans. viii., 1818, pp. 157-164.

31. On a new Optical and Mineralogical Property of Calcareous Spar (1816).—Edinb. Roy. Soc. Trans. viii., 1818, pp. 165-170.

32. On the effects of Compression and Dilatation in altering the Polarizing structure of Doubly Refracting Crystals (1816).—Edinb. Roy. Soc. Trans. viii., 1818, pp. 281-286.

33. On the Laws which regulate the distribution of the Polarizing Force in Plates, Tubes, and Cylinders of Glass that have received the Polarizing Structure (1816).—Edinb. Roy. Soc. Trans. viii., 1818, pp. 353-372.

34. On the Laws of Polarization and Double Refraction in regularly Crystallized Bodies.—Phil. Trans., 1818, pp. 199-272.

35. Description of a method of making Doubly Refracting Prisms perfectly Achromatic.—Thomson, Ann. Phil. xi., 1818, pp. 175-177.

36. On the difference between the Optical Properties of Arragonite and Calcareous Spar.—Quart. Journ. Sci. iv., 1818, pp. 112-114.

37. Optical structure of Ice.—Quart. Journ. Sci. iv., 1818, p. 155.

38. On a new Optical and Mineralogical Structure, exhibited in certain Specimens of Apophyllite and other Minerals.—Edinb. Phil. Journ. i., 1819, pp. 1-8 ; Edinb. Roy. Soc. Trans. ix., 1823, pp. 317-336.

39. On the Phosphorescence of Minerals.—Edinb. Phil. Journ. i., 1819, pp. 383-388.

40. On the Laws which regulate the Absorption of Polarized Light by Doubly Refracting Crystals (1818).—Phil. Trans., 1819, pp. 11-28 ; Edinb. Phil. Journ. ii., 1820, pp. 341-348.

41. On the action of Crystallized Surfaces upon Light.—Phil. Trans., 1819, pp. 145-160.

42. On the Optical and Physical Properties of Tabasheer.—Phil. Trans., 1819, pp. 283-299 ; Edinb. Phil. Journ. i., 1819, pp. 147-150 ; ii., 1820, pp. 97-102.

43. On a singular Development of Crystalline Structure by Phosphorescence.—Edinb. Phil. Journ. ii., 1820, pp. 171, 172.

44. On the Optical Properties and Mechanical Condition of Amber.—Edinb. Phil. Journ. ii., 1820, pp. 332-334.

45. Account of some Single Microscopes upon a new construction.—Edinb. Phil. Journ. iii., 1820, pp. 74-77.

46. Notice respecting a singular Structure in the Diamond.—Edinb. Phil. Journ. iii., 1820, pp. 98-100.

47. Notice respecting some new species of Lead-Ore from Wanlochhead and Leadhills.—Edinb. Phil. Journ. iii., 1820, pp. 138-140.

48. On the Phenomena of Dichroism, or the Absorption of Common Light by Crystallized Bodies.—Edinb. Phil. Journ. iii., 1820, pp. 243-247.

49. On a singular Luminous Property of Wood, etc., steeped in Solutions of Lime and Magnesia.—Edinb. Phil. Journ. iii., 1820, pp. 343-344.

50. Additions aux Observations sur les rapports entre la forme Primitive des Minéraux et le Nombre de leurs axes de Double Réfraction.—Journ. de Phys. xci., 1820, pp. 300-309.

51. Account of Comptonite, a new Mineral from Vesuvius.—Edinb. Phil. Journ. iv., 1821, pp. 131-133.

52. Description of a new Double Image Micrometer for Measur-

ing the Distance of Celestial Objects.—Edinb. Phil. Journ. iv., 1821, pp. 164-167.

53. Reply to a Note in the "Annales de Chimie," by M. Arago, on the Phosphorescence of Fluor-Spar.—Edinb. Phil. Journ. iv., 1821, pp. 180-185.

54. Account of the new Hydrate of Magnesia discovered in Shetland.—Edinb. Phil. Journ. iv., 1821, pp. 352-355 ; Edinb. Roy. Soc. Trans. ix., 1823, pp. 239-242.

55. Account of the Atush-Kudda, or Natural Fire Temples of the Guebres, formed by Burning Springs of Naphtha, with a Notice respecting the Naphtha Wells in Pegu.—Edinb. Phil. Journ. v., 1821, pp. 21-27.

56. On the Connexion between the Optical Structure and Chemical Composition of Minerals.—Edinb. Phil. Journ. v., 1821, pp. 1-8.

57. On the Form of the Integrant Molecule of Carbonate of Lime (1818).—Geol. Soc. Trans. v., 1821, pp. 83-86.

58. Observations on Vision through Coloured Glasses, and on their Application to Telescopes and Microscopes of great magnitude.— Edinb. Phil. Journ. vi., 1822, pp. 102-107.

59. Description of a Teinoscope for altering the Lineal Proportions of Objects ; with Observations on Professor Amici's Memoir on Telescopes without Lenses.—Edinb. Phil. Journ. vi., 1822, pp. 334-338.

60. Observations on the Relation between the Optical Structure and the Chemical Composition of the Apophyllite, and other Minerals of the Zeolite Family, in reference to the Analyses of M. Berzelius.—Edinb. Phil. Journ. vii., 1822, pp. 12-18.

61. Account of a singular Experiment depending on the Polarization of Light by Reflection.—Edinb. Phil. Journ. vii., 1822, pp. 146, 147.

62. Description of a new Reflecting Telescope.—Edinb. Phil. Journ. vii., 1822, pp. 323-328 ; viii., 1823, pp. 326, 327.

63. On the Construction of Polyzonal Lenses and Mirrors of great magnitude for Lighthouses and for Burning Instruments, and on the Formation of a great National Burning Apparatus.—Edinb. Phil. Journ. viii., 1823, pp. 160-169 ; Edinb. Roy. Soc. Trans. xi., 1831, pp. 33-72.

64. Description of a new Reflecting Microscope.—Edinb. Phil. Journ. viii., 1823, pp. 326, 327.

65. On the Existence of two new Fluids in the Cavities of Minerals, which are Immiscible, and possess remarkable Physical Properties.—Edinb. Phil. Journ. ix., 1823, pp. 94, 95, 268-270, 400 ; Edinb. Roy. Soc. Trans. x., 1826, pp. 1-42.

66. On the Knights' Moves over the Chess-board.—Edinb. Phil. Journ. ix., 1823, pp. 236, 237.

67. Reply to Mr. Brooke's Observations on the Connexion between the Optical Structure of Minerals and their Primitive Forms. —Edinb. Phil. Journ. ix., 1823, pp. 361-372.

68. On Circular Polarization, as exhibited in the Optical Structure of the Amethyst, with remarks on the Distribution of the Colouring Matter in that Mineral (1819).—Edinb. Roy. Soc. Trans. ix., 1823, pp. 139-152.

69. Observations on the Mean Temperature of the Globe (1820). —Edinb. Roy. Soc. Trans. ix., 1823, pp. 201-226 ; Edinb. Journ. Sci. iv., 1831, pp. 300-320.

70. Description of a Monochromatic Lamp for Microscopical Purposes, etc., with remarks on the Absorption of the Prismatic Rays by Coloured Media (1822).—Edinb. Roy. Soc. Trans. ix., 1823, pp. 433-444.

71. On a new Species of Double Refraction, accompanying a remarkable structure in Analcine (1822).— Edinb. Phil. Journ. x., 1824, pp. 255-259 ; Edinb. Roy. Soc. Trans. x., 1826, pp. 187-194.

72. On the Accommodation of the Eye to different distances (1823).—Edinb. Journ. Sci. i., 1824, pp. 77-83.

73. Historical account of the Discoveries respecting the Double Refraction and Polarization of Light.—Edinb. Journ. Sci. i., 1824, pp. 90-96.

74. Description of Two Surfaces composed of Siliceous Filaments incapable of reflecting Light, and produced by the fracture of a large piece of Quartz.—Edinb. Journ. Sci. i., 1824, pp. 108-110.

75. Observations on the Pyro-Electricity of Minerals.—Edinb. Journ. Sci. i., 1824, pp. 208-215.

76. Observations of the Vision of Impressions on the Retina, in reference to certain supposed discoveries respecting Vision announced by Mr. Charles Bell.—Edinb. Journ. Sci. ii., 1825, pp. 1-9.

77. On the formation of Single Microscopes from the Lenses of Fishes, etc.—Edinb. Journ. Sci. ii., 1825, pp. 98-100.

78. Description of an extraordinary Parhelion observed at Gotha, 12th May 1824.—Edinb. Journ. Sci. ii., 1825, pp. 105-108.

79. On the Structure of Rice Paper (1822).—Edinb. Journ. Sci. ii., 1825, pp. 135-136.

80. On the Optical Structure of the Lithion-Mica, analysed by Professor Gmelin.—Edinb. Journ. Sci. ii., 1825, pp. 205, 206.

81. Description of Withamite, a new Mineral Species found in Glencoe.—Edinb. Journ. Sci. ii., 1825, pp. 218-224.

82. Description of Gmelinite, a new Mineral Species.—Edinb. Journ. Sci. ii., 1825, pp. 262-267.

83. A description of Levyne, a new Mineral Species.—Edinb. Journ. Sci. ii., 1825, pp. 332-334; iv., 1826, pp. 316, 317.

84. On some remarkable Affections of the Retina as exhibited in its Insensibility to Indirect Impressions, and to the Impressions of Attenuated Light (1822).—Edinb. Journ. Sci. iii., 1825, pp. 288-293 ; Brit. Assoc. Rep., 1831-32, pp. 551-553.

85. On the Optical Illusion of the Conversion of Cameos into Intaglios and of Intaglios into Cameos, with an account of other Analogous Phenomena.—Edinb. Journ. Sci. iv., 1826, pp. 99-108.

86. Results of the Thermometrical Observations made at Leith Fort every hour of the day and night, during the whole of the years 1824 and 1825.—Edinb. Roy. Soc. Trans. x., 1826, pp. 362-388 ; Edinb. Journ. Sci. v., 1826, pp. 18-32.

87. On the Refractive Power of the two new Fluids in Minerals, with additional Observations on the Nature and Properties of these Substances.—Edinb. Roy. Soc. Trans. x., 1826, pp. 407-427 ; Edinb. Journ. Sci. v., 1826, pp. 122-136.

88. Description of Hoperite, a new Mineral from Altenberg near Aix-la-Chapelle (1823).—Edinb. Roy. Soc. Trans. x., 1826, pp. 107-111.

89. On the Distribution of the Colouring Matter, and on certain Peculiarities in the Structure and Properties of the Brazilian Topaz (1822).—Cambridge Phil. Soc. Trans. ii., 1827, pp. 1-10.

90. Notice respecting the Mean Temperature of the Equator.— Edinb. Journ. Sci. vi., 1827, pp. 117-120.

91. Notice respecting the Existence of the new Fluid in a Large Cavity in a specimen of Sapphire.—Edinb. Journ. Sci. vi., 1827, pp. 155, 156.

92. On the separation of Epistilbite from Heulandite, as demon-

strated by Optical Characters.—Edinb. Journ. Sci. vi., 1827, pp. 236-240.

93. Account of an Improvement in the Nautical Eye-Tube.— Edinb. Journ. Sci. vi., 1827, pp. 250, 251.

94. Observations on the Structure and Crystalline Forms of Hayterite.—Edinb. Journ. Sci. vi., 1827, pp. 301-307.

95. On the Systems of Double Stars, which have been demonstrated to be Binary ones by the observations of Sir W. Herschel and Messrs. Herschel and South.—Edinb. Journ. Sci. vii., 1827, pp. 88-97 ; viii., 1828, pp. 40-44.

96. Description of Oxahverite, a new Mineral from Oxahver, in Iceland.—Edin. Journ. Sci. vii., 1827, pp. 115-118.

97. Notice respecting Professor Barlow's new Achromatic Telescopes with Fluid Object-Glasses.—Edinb. Journ. Sci. vii., 1827, pp. 335, 336.

98. A Table of Refractive Densities.—Quart. Journ. Sci. xxii., 1827, pp. 355-365.

99. Account of the Fossil Bones found on the left bank of the Irrawadi in Ava.—Edinb. Journ. Sci. viii., 1828, pp. 56-60.

100. On the Mean Temperature of the Equator as deduced from observations made at Prince of Wales' Island, Singapore, and Malacca.—Edinb. Journ. Sci. viii., 1828, pp. 60-67.

101. Notice respecting the Varnish and Varnish Trees of India. —Edinb. Journ. Sci. viii., 1828, pp. 96-100.

102. Account of the Poisonous qualities of the Vegetable Varnishes of America and India.—Edinb. Journ. Sci. viii., 1828, pp. 100-104.

103. On the supposed influence of the Aurora Borealis upon the Magnetic Needle, in reply to the Observations of M. Arago, as communicated to the Academy of Sciences, Jan. 22, 1828.—Edinb. Journ. Sci. viii., 1828, pp. 189-201.

104. On the Natural History and Properties of Tabasheer, the Siliceous Concretion in the Bamboo.—Edinb. Journ. Sci. viii., 1828, pp. 285-294.

105. Account of the Performances of different Ventriloquists, with observations on the Art of Ventriloquism.—Edinb. Journ. Sci. ix., 1828, pp. 252-259.

106. On a new Cleavage in Calcareous Spar, with a Notice of a Method of Detecting Secondary Cleavages in Minerals.—Edinb. Journ. Sci. ix., 1828, pp. 311-314.

107. Facts and Observations relative to the Recent Formation of Quartz Crystals, etc., and of Indurated Calcedony from Siliceous Solutions and Paste.—Edinb. Journ. Sci. x., 1829, pp. 28-33.

108. Account of Two remarkable Cases of Colour-Blindness.—Edinb. Journ. Sci. x., 1829, pp. 153-155.

109. Account of Two remarkable Rainbows, one of which enclosed the Phenomenon of Converging Solar Beams.—Edinb. Journ. Sci. x., 1829, pp. 163, 164.

110. Observations relative to the Motions of the Molecules of Bodies (1828).—Edinb. Journ. Sci. x., 1829, pp. 215-220.

111. Account of a remarkable Peculiarity in the Structure of Glauberite, which has one axis of double Refraction for Violet, and two axes for Red Light (1828).—Edinb. Journ. Sci. x., 1829, pp. 325-327.

112. Account of the Single Lens Microscope of Sapphire and Diamond, constructed by Mr. A. Pritchard.—Edinb. Journ. Sci. x., 1829, pp. 327-333.

113. Notice respecting a Method of Producing an Intense Heat from Gas for various purposes in the Arts (1826).—Edinb. Journ. Sci. i., 1829, pp. 104-108.

114. Account of a new Monochromatic Lamp, depending on the Combustion of Compressed Gas (1826).—Edinb. Journ. Sci. i., 1829, p. 108.

115. On the Reflexion and Decomposition of Light at the Separating Surfaces of Media of the same, and of different Refractive Powers.—Phil. Trans., 1829, pp. 187-206 ; Edinb. Journ. Sci. i., 1829, pp. 209-229.

116. On a new Series of Periodical Colours produced by the Grooved Surfaces of Metallic and Transparent Bodies.—Phil. Trans., 1829, pp. 301-316 ; Edinb. Journ. Sci. ii., 1830, pp. 46-61.

117. On the Law of the partial Polarization of Light by Reflexion (1829).—Edinb. Journ. Sci. iii., 1830, pp. 160-177.

118. Microscopical Examination of the Structure of the Second Stomach of certain Cetacea.—Edinb. Journal of Science, iii., 1830, p. 325.

119. On the Production of Regular Double Refraction in the Molecules of Bodies by Simple Pressure, with Observations on the Origin of the Doubly Refracting Structure.—Phil. Trans., 1830, pp. 87-96 ; Edinb. Journ. Sci. iii., 1830, pp. 328-337.

120. On the Laws of the Polarization of Light by Refraction

(1829).—Phil. Trans., 1830, pp. 69-84, 133-144 ; Edinb. Journ. Sci. iii., 1830, pp. 218-230.

121. On the action of the Second Surfaces of Transparent Plates upon Light.—Phil. Trans., 1830, pp. 145-152; Edinb. Journ. Sci. iii., 1830, pp. 230-237.

122. On the Phenomena and Laws of Elliptic Polarization as exhibited in the action of Metals upon Light.—Phil. Trans., 1830, pp. 287-326 ; Edinb. Journ. Sci. iv., 1831, pp. 136-165, 247-261.

123. Remarks on Dr. Goring's Observations on the Use of Mono-chromatic Light with the Microscope.—Edinb. Journ. Sci. v., 1831, pp. 143-148.

124. On a new Analysis of Solar Light indicating three Primary Colours, forming Coincident Spectra of equal Length.—Edinb. Journ. Sci. v., 1831, pp. 197-206 ; Edinb. Roy. Soc. Trans. xii., 1834, pp. 123-136.

125. On the Mean Temperature of Thirty-four Different Places in the State of New York for 1830.—Edinb. Journ. Sci. v., 1831, pp. 255-265.

126. On the Construction of Polyzonal Lenses and their Com-bination with Plain Mirrors, for the purposes of Illumination in Lighthouses (1827).—Edinb. Roy. Soc. Trans. xi., 1831, pp. 33-72.

127. On certain new Phenomena of Colour in Labrador Felspar, with Observations on the Nature and Cause of its changeable Tints (1829).—Edinb. Roy. Soc. Trans. xi., 1831, pp. 322-331.

128. Ueber die mathematischen Ausdrücke für die Mittlere Wäre der Luft und die Magnetische Intensität.—Poggend. Annal. xxi. 1831, pp. 323-325.

129. On the Structure of the Crystalline Lens in Fishes, Birds, Reptiles, and Quadrupeds.—Brit. Assoc. Rep., 1831-32, pp. 81, 82.

130. Report on the Recent Progress of Optics.—Brit. Assoc. Rep., 1831-32, pp. 308-322.

131. On the Undulations excited in the Retina by the action of Luminous Points and Lines.—Brit. Assoc. Rep., 1831-32, pp. 548-551.

132. On the Principle of Illumination for Microscopic Objects (1831).—Edinb. Journ. Sci. vi., 1832, pp. 83-85.

133. Observations on M. Rudberg's Memoir on Double Refraction —Phil. Mag. i., 1832, p. 6.

134. On a new Species of Coloured Fringes produced from Re-

flection between the Lenses of Achromatic Compound Object-Glasses.—Phil. Mag. i., 1832, pp. 19-23 ; Edinb. Roy. Soc. Trans. xii., 1834, pp. 191-196.

135. On the effect of Compression and Dilatation upon the Retina.—Phil. Mag. i., 1832, pp. 89-92.

136. On his Formulæ for Mean Temperature.—Phil. Mag. i., 1832, p. 135.

137. On the action of Heat in changing the Number and Nature of the Optical or Resultant Axes of Glauberite.—Phil. Mag. i., 1832, pp. 417-420.

138. Account of a curious Chinese Mirror which reflects from its polished face the Figures embossed upon its back.—Phil. Mag. i., 1832, pp. 438-441.

139. Observations on the Isothermal Lines on the North-west Coast of America as deduced from Rupffer's Observations.—Phil. Mag. i., 1832, pp. 431, 432.

140. Observations on the Action of Light upon the Retina.— Phil. Mag. ii., 1833, pp. 168-175.

141. On the Absorption of Specific Rays in reference to the Undulatory Theory of Light.—Phil. Mag. ii., 1833, pp. 360-363.

142. Notice respecting certain Changes of Colour in the Choroid Coat of the Eye of Animals.—Phil. Mag. iii., 1833, pp. 288, 289.

143. On the Anatomical and Optical Structure of the Crystalline Lenses of Animals, particularly of the Cod.—Phil. Trans., 1833, pp. 323-332 ; Phil. Mag. viii., 1836, pp. 193-202.

144. Observations on the Lines of the Solar Spectrum, and on those produced by the Earth's Atmosphere, and by the action of Nitrous Acid Gas (1833).— Edinb. Roy. Soc. Trans. xii., 1834, pp. 519-530 ; Phil. Mag. viii., 1836, pp. 384-392.

145. On the Colours of Natural Bodies.—Edinb. Roy. Soc. Trans. xii., 1834, pp. 538-545.

146. On the supposed Vision of the Blood-vessels of the Eye.— Phil. Mag. iv., 1834, pp. 115-120.

147. On the Influence of Successive Impulses of Light upon the Retina.—Phil. Mag. iv., 1834, pp. 241-245.

148. Account of Two Experiments on accidental Colours, with Observations on their Theory.—Phil. Mag. iv., 1834, pp. 353-354.

149. Notice respecting a remarkable Specimen of Amber.—Brit. Assoc. Rep., 1834, pp. 574, 575.

150. Remarks on the value of Optical Characters in the discrimination of Mineral Species.—Brit. Assoc. Rep., 1834, p. 575.

151. Observations relative to the Structure and Origin of the Diamond (1833).—Geol. Soc. Trans. iii., 1835, pp. 455-460 ; Phil. Mag. vii., 1835, pp. 245-250.

152. On the supposed Achromatism of the Eye.—Phil. Mag. vi., 1835, pp. 161-164.

153. On certain Peculiarities in the Double Refraction and Absorption of Light exhibited in the Oxalate of Chromium and Potash.—Phil. Trans., 1835, pp. 91-94 ; Phil. Mag., 1835, pp. 436-439.

154. On the action of Crystallized Surfaces upon Common and Polarized Light.—Brit. Assoc. Rep., 1836 (Pt. 2), pp. 13-16 ; Phil. Mag. xii., 1838, pp. 22-27.

155. On a singular Development of Polarizing Structure in the Crystalline Lens after Death.—Brit. Assoc. Rep., 1836 (Pt. 2), pp. 16-18 ; Phil. Mag. xii., 1838, pp. 22-27.

156. On the Connexion between the Phenomena of the Absorption of Light and the Colours of Thin Plates.—Phil. Trans., 1837, pp. 245-252 ; Phil. Mag. xxi., 1842, pp. 208-217.

157. On the cause of the Optical Phenomena which take place in the Crystalline Lens during the Absorption of Distilled Water.—Brit. Assoc. Rep., 1837 (Pt. 2), pp. 11, 12.

158. On a new Property of Light.—Brit. Assoc. Rep., 1837 (Pt. 2), pp. 12, 13.

159. Notice of a new Structure in the Diamond.—Brit. Assoc. Rep., 1837 (Pt. 2), pp. 13-15.

160. On an Ocular Parallax in Vision, and on the Law of Visible Direction.—Brit. Assoc. Rep., 1838 (Pt. 2), pp. 7-9.

161. On a new Phenomenon of Colour in certain Specimens of Fluor-Spar.—Brit. Assoc. Rep., 1838 (Pt. 2), pp. 10-12.

162. An account of certain new Phenomena of Diffraction.—Brit. Assoc. Rep., 1838 (Pt. 2), p. 12.

163. On the combined action of Grooved Metallic and Transparent Surfaces upon Light.—Brit. Assoc. Rep., 1838 (Pt. 2), p. 13.

164. On a new kind of Polarity in Homogeneous Light.—Brit. Assoc. Rep., 1838 (Pt. 2), pp. 13, 14.

165. On some preparations of the Eye by Mr. Clay Wallace of New York.—Brit. Assoc. Rep., 1838 (Pt. 2), pp. 14, 15.

166. On the Structure of the Fossil Teeth of the Sauroid Fishes. —Brit. Assoc. Rep., 1838 (Pt. 2), pp. 90, 91.

167. Analyse des Gmelinits oder Hydroliths.—Liebig, Annal. xxviii., 1838, p. 342.

168. On the colours of Mixed Plates (1837).—Phil. Trans., 1838, pp. 73-78 ; Phil. Mag. xiv., 1839, pp. 191-196.

169. Report respecting the two series of Hourly Meteorological Observations kept in Scotland.—Brit. Assoc. Rep., 1839, pp. 27-29.

170. Observations on Professor Plateau's Defence of his Theory of Accidental Colours.—Phil. Mag. xv., 1839, pp. 435-441.

171. On Professor Powell's Measures of the Indices of Refraction for the lines G and H in the Spectrum.—Brit. Assoc. Rep., 1840 (Pt. 2), p. 5.

172. On the Decomposition of Glass.—Brit. Assoc. Rep., 1840 (Pt. 2), pp. 5, 6.

173. On the Rings of Polarized Light produced in Specimens of Decomposed Glass.—Brit. Assoc. Rep., 1840 (Pt. 2), pp. 6, 7.

174. On the cause of the Increase of Colour by the Inversion of the Head.—Brit. Assoc. Rep., 1840 (Pt. 2), pp. 7, 8.

175. On the Phenomena and Cause of Muscæ Volitantes.—Brit. Assoc. Rep., 1840 (Pt. 2), pp. 8, 9.

176. On a Method of Illuminating Microscopic Objects.—Brit. Assoc. Rep., 1840 (Pt. 2), pp. 9, 10.

177. On an Improvement in the Polarizing Microscopes.—Brit. Assoc. Rep., 1840 (Pt. 2), p. 10.

178. On a remarkable Rainbow observed by Mr. Bowman.—Brit. Assoc. Rep., 1840 (Pt. 2), p. 12.

179. Report respecting the two series of Hourly Meteorological Observations kept at Inverness and Kingussie from November 1st, 1838, to November 1st, 1839.—Brit. Assoc. Rep., 1840, pp. 349-352.

180. On the Optical Figures produced by the Disintegrated Surfaces of Crystals (1837).— Edinb. Roy. Soc. Trans. xiv., 1840, pp. 164-175 ; Phil. Mag. v., 1853, pp. 16-28.

181. On a remarkable Property of the Diamond.—Phil. Trans. 1841, pp. 41, 42.

182. On the Phenomena of Thin Plates of Solid and Fluid Substances exposed to Polarized Light.—Phil. Trans., 1841, pp. 43-58 ; Phil. Mag. xxxii., 1848, pp. 181-199.

183. On the Compensations of Polarized Light, with the description of a Polarimeter for measuring degrees of Polarization.—Roy. Soc. Proc. iv., 1841, pp. 306, 307 ; Irish Acad. Trans. xix., 1843, pp. 377-392.

184. On the Colouring Matter of certain Land-shells from the Philippine Islands.—Zool. Soc. Proc. ix., 1841, pp. 15, 16.

185. On a new Property of the Rays of the Spectrum, with Observations on the explanation of it given by the Astronomer-Royal on the Principles of the Undulatory Theory.—Brit. Assoc. Rep., 1842 (Pt. 2), p. 12.

186. On the Dichroism of the Palladio-chlorides of Potassium and Ammonium.—Brit. Assoc. Rep., 1842 (Pt. 2), p. 13.

187. On the existence of a new Neutral Point and Two Secondary Neutral Points.—Brit. Assoc. Rep., 1842 (Pt. 2), p. 13.

188. On Crystalline Reflection.—Brit. Assoc. Rep., 1842 (Pt. 2), pp. 13, 14.

189. On a very curious fact connected with Photography, discovered by Mr. Möser of Königsberg.—Brit. Assoc. Rep., 1842 (Pt. 2), p. 14.

190. On the Geometric Forms and Laws of Illumination on the Spaces which receive the Solar Rays transmitted through Quadrangular Apertures.—Brit. Assoc. Rep., 1842 (Pt. 2), p. 15.

191. On Luminous Lines in certain Flames corresponding to the Defective Lines in the Sun's Light.—Brit. Assoc. Rep., 1842 (Pt. 2), p. 15.

192. On the Structure of a part of the Solar Spectrum hitherto unexamined.—Brit. Assoc. Rep., 1842 (Pt. 2), p. 15.

193. On the Luminous Bands in the Spectra of various Flames.—Brit. Assoc. Rep., 1842 (Pt. 2), pp. 15, 16.

194. Notice of an Experiment on the Ordinary Refraction of Iceland Spar (with remarks by Mac Cullagh).—Brit. Assoc. Rep., 1843 (Pt. 2), pp. 7, 8.

195. On the action of Two Blue Oils upon Light.—Brit. Assoc. Rep., 1843 (Pt. 2), p. 8.

196. On the Cause of the Colours in Iridescent Agate.—Phil. Mag. xxii., 1843, pp. 213-215.

197. On the Combination of prolonged, direct, luminous Impressions on the Retina, with their complementary Impressions.—Phil. Mag. xxii., 1843, pp. 434, 435.

198. A Notice explaining the Cause of an Optical Phenomenon observed by the Rev. W. Selwyn.—Brit. Assoc. Rep., 1844 (Pt. 2), p. 8.

199. An account of the Cause of Colours in the precious Opal.—Brit. Assoc. Rep., 1844 (Pt. 2), p. 9.

200. Notice respecting the Cause of beautiful White Rings which are seen round a Luminous Body when looked at through certain specimens of Calcareous Spar.—Brit. Assoc. Rep., 1844 (Pt. 2), p. 9.

201. On Crystals in the Cavities of Topaz, which are dissolved by heat, and Re-crystallize on cooling.—Brit. Assoc. Rep., 1844 (Pt. 2), pp. 9, 10.

202. On the Accommodation of the Eye to various distances.—Brit. Assoc. Rep., 1844 (Pt. 2), pp. 10, 11.

203. Account of a series of Experiments on the Polarization of Light by Rough Surfaces and White Dispersing Surfaces.—Brit. Assoc. Rep., 1844 (Pt. 2), p. 11.

204. On the Law of Visible Position in Single and Binocular Vision, and on the Representation of Solid Figures by the Union of dissimilar Plane Pictures on the Retina (1843).—Edinb. Roy. Soc. Trans. xv., 1844, pp. 349-368.

205. On the Optical Phenomena, Nature, and locality of Muscæ Volitantes, with Observations on the Structure of the Vitreous Humour, and on the Vision of Objects placed within the Eye (1843). —Edinb. Roy. Soc. Trans. xv., 1844, pp. 377-386.

206. On the Conversion of Relief by Inverted Vision.—Edinb. Roy. Soc. Trans. xv., 1844, pp. 657-662 ; Phil. Mag. xxx., 1847, pp. 432-437.

207. On the Knowledge of Distance given by Binocular Vision.—Edinb. Roy. Soc. Trans. xv., 1844, pp. 663-675.

208. Observations on Colour-Blindness, or Insensibility to the Impressions of certain Colours.—Phil. Mag. xxv., 1844, pp. 134-141.

209. On a new Polarity of Light, with an examination of Mr. Airy's explanation of it on the Undulatory Theory.—Brit. Assoc. Rep., 1845 (Pt. 2), pp. 7, 8.

210. Notice of Two New Properties of the Retina.—Brit. Assoc. Rep., 1845 (Pt. 2), pp. 8, 9.

211. An Improvement in the Method of taking Positive Talbo-types (Calotypes).—Brit. Assoc. Rep., 1845 (Pt. 2), pp. 10, 11.

212. On Fog-rings observed in America.—Brit. Assoc. Rep., 1845 (Pt. 2), p. 19.

213. On the Optical Properties of Greenockite (1843).—Edinb. Roy. Soc. Proc. i., 1845, pp. 342, 343.

214. On certain Negative Actions of Light.—Edinb. Roy. Soc. Proc. i., 1845, pp. 422-424.

215. Sur la Polarization de la Lumière Atmosphérique.—Paris, Comptes Rendus, xx., 1845, pp. 801, 802.

216. On the Discovery of the Composition of Water.—Phil. Mag. xxvii., 1845, pp. 195-197.

217. Notice of a new Property of Light exhibited in the action of Chrysammate of Potash upon Common and Polarized Light.—Brit. Assoc. Rep., 1846 (Pt. 2), pp. 7-8.

218. On the Diffraction Bands produced by the Edges of Thin Plates, whether Solid or Fluid.—Brit. Assoc. Rep., 1847 (Pt. 2), p. 33.

219. On the Dark Lines in the portion of the Red Space beyond the Red Extremity of the Spectrum, as seen by Fraunhöfer.—Brit. Assoc. Rep., 1847 (Pt. 2), p. 33.

220. An account of the Functions of the Membranes of the Eye at the Foramen Centrale of Soemmering.—Brit. Assoc. Rep., 1847 (Pt. 2), p. 33.

221. On the New Analysis of Solar Light.—Phil. Mag. xxx., 1847, pp. 153-158.

222. On the Knowledge of Distance given by Binocular Vision. —Phil. Mag. xxx., 1847, pp. 305-318.

223. On the Analysis of the Spectrum by Absorption.—Phil. Mag. xxx., 1847, pp. 461, 462.

224. On the Modification of the Doubly Refracting and Physical Structure of Topaz, by Elastic Forces emanating from minute Cavities.—Phil. Mag. xxxi., 1847, pp. 101-104 ; Edinb. Roy. Soc. Trans. xvi., 1849, pp. 7-10.

225. On the Polarization of the Atmosphere.—Phil. Mag. xxxi., 1847, pp. 444-454.

226. On Crystals in the Cavities of Minerals.—Phil. Mag. xxxi., 1847, pp. 497-510.

227. On the Compensation of Impressions moving over the Retina, as seen in Railway Travelling.—Brit. Assoc. Rep., 1848 (Pt. 2), pp. 47, 48.

APPENDIX. 433

228. On the Vision of Distance, as given by Colour.—Brit. Assoc. Rep., 1848 (Pt. 2), p. 48.

229. On the Visual Impressions upon the Foramen Centrale of the Retina.—Brit. Assoc. Rep., 1848 (Pt. 2), pp. 48, 49.

230. An Examination of Bishop Berkeley's "New Theory of Vision."—Brit. Assoc. Rep., 1848 (Pt. 2), p. 49.

231. On the Decomposition and Dispersion of Light within Solid and Fluid Bodies.—Phil. Mag. xxx., 1848, pp. 401-412.

232. On the Optical Phenomena, Nature, and Locality of Muscæ Volitantes.—Phil. Mag. xxxii., 1848, pp. 1-11.

233. On the Distinctness of Vision produced in certain cases by the Use of the Polarizing Apparatus in Microscopes.—Phil. Mag. xxxii., 1848, pp. 161-165.

234. On the Elementary Colours of the Spectrum, in the Reply to M. Melloni.—Phil. Mag. xxxii., 1848, pp. 489-494.

235. On the Phenomenon of Luminous Rings in Calcareous Spar and Beryl, as produced by Cavities containing the Two new Fluids. —Phil. Mag. xxxiii., 1848, pp. 489-493.

236. Description of a Binocular Camera.—Brit. Assoc. Rep., 1849 (Pt. 2), p. 5.

237. Improvement of the Photographic Camera.—Brit. Assoc. Rep., 1849 (Pt. 2), p. 5.

238. On a new Form of Lenses, and their Application to the Construction of Two Telescopes or Microscopes of exactly equal optical power.—Brit. Assoc. Rep., 1849 (Pt. 2), p. 6.

239. Notice of Experiments on Circular Crystals.—Brit. Assoc. Rep. 1849 (Pt. 2), p. 6.

240. An account of a new Stereoscope.—Brit. Assoc. Rep., 1849 (Pt. 2), pp. 6, 7 ; Edinb. New Phil. Journ. xlviii., 1850, pp. 150, 151.

241. On the Existence of Crystals with different Primitive Forms and Physical Properties in the Cavities of Minerals ; with Additional Observations on the new Fluids in which they occur (1845). —Edinb. Roy. Soc. Trans. xvi., 1849, pp. 11-22.

242. On the Decomposition and Dispersion of Light within Solid and Fluid Bodies (1846).—Edinb. Roy. Soc. Trans. xvi., 1849, pp. 111-122.

243. Notice on the artificial Magnets made by M. Logeman by the Process of M. Elias.—Brit. Assoc. Rep., 1850 (Pt. 2), p. 4.

244. On a new Membrane investing the Crystalline Lens of the Ox.—Brit. Assoc. Rep., 1850 (Pt. 2), pp. 4, 5.

245. On the Optical Properties of the Cyanurets of Platinum and Magnesia, and of Barytes and Platinum.—Brit. Assoc. Rep., 1850 (Pt. 2), p. 5.

246. On the Polarizing Structure of the Eye.—Brit. Assoc. Rep., 1850 (Pt. 2), pp. 5, 6.

247. On some new Phenomena in the Polarization of the Atmosphere.—Brit. Assoc. Rep. 1850 (Pt. 2), p. 6.

248. Sur quelques Phénomènes de Polarisation qui ont rapport avec la Diffraction opérée par les Surfaces rayées.—Paris, Comptes Rendus, xxx., 1850, pp. 496-498.

249 Observations sur les points neutres de l'Atmosphère découverts par MM. Arago et Babinet.—Paris, Comptes Rendus, xxx., 1850, pp. 532-536.

250. Observations sur le Spectre solaire.—Paris, Comptes Rendus, xxx., 1850, pp. 578-581.

251. Description of several new and simple Stereoscopes for exhibiting as Solids one or more representations of the Solid on a Plane.—Edinb. Trans. Scot. Soc. Arts, iii., 1851, pp. 247-258 ; Phil. Mag. iii., 1852, pp. 16-26.

252. Account of a Binocular Camera, and of a Method of obtaining drawings of full-length and colossal statues, and of living bodies, which can be exhibited as Solids by the Stereoscope.—Edinb. Trans. Scot. Soc. Arts, iii., 1851, pp. 259-264; Phil. Mag. iii., 1852, pp. 26-31.

253. Notice of a Tree struck by Lightning in Clandeboye Park.—Brit. Assoc. Rep., 1852 (Pt. 2), pp. 2, 3.

254. Account of a Case of Vision without Retina.—Brit. Assoc. Rep., 1852 (Pt. 2), p. 3.

255. On the Form of Images produced by Lenses and Mirrors of different sizes.—Brit. Assoc. Rep., 1852 (Pt. 2), pp. 3-6.

256. Notice of a Chromatic Stereoscope.—Phil. Mag. iii., 1852, pp. 31, 32.

257. Explanation of an Optical Illusion.—Phil. Mag. iii., 1852, pp. 55-57.

258. On the Development and Extinction of regular Doubly Refracting Structures in the Crystalline Lenses of Animals after Death.—Phil. Mag. iii., 1852, pp. 192-198.

259. On the Optical Phenomena and Crystallization of Tourmaline, Titanium, and Quartz within Mica, Amethyst, and Topaz.—Edinb. Roy. Soc. Trans. xx., 1853, pp. 547-554; Phil. Mag. vi., 1853, 265-272.

260. On the Production of Crystalline Structure in Crystallized Powders by Compression and Traction.—Edinb. Roy. Soc. Trans. xx., 1853, pp. 555-560 ; Phil. Mag. vi., 1853, pp. 260-264.

261. On Circular Crystals.—Edinb. Roy. Soc. Trans. xx., 1853, pp. 607-624.

262. On Cavities in Amber containing Gases and Fluids.—Phil. Mag. v., 1853, pp. 233-235.

263. Account of a remarkable Fluid Cavity in Topaz.—Phil. Mag. v., 1853, pp. 235, 236.

264. Notice on Barometrical, Thermometrical, and Hygrometrical Clocks.—Phil. Mag. vii., 1854, p. 358.

265. On the Binocular Vision of Surfaces of different Colours.—Brit. Assoc. Rep., 1855 (Pt. 2), p. 9.

266. On the Triple Spectrum.—Brit. Assoc. Rep., 1855 (Pt. 2), pp. 7-9.

267. On the Existence of Acari in Mica.—Brit. Assoc. Rep., 1855 (Pt. 2), p. 9.

268. On the Absorption of Matter by the Surfaces of Bodies.—Brit. Assoc. Rep., 1855 (Pt. 2), p. 9.

269. On the Centering of the Lenses of the Compound Object-Glasses of Microscopes.—Brit. Assoc. Rep., 1857 (Pt. 2), pp. 4, 5.

270. On the Duration of Luminous Impressions on certain Points of the Retina.—Brit. Assoc. Rep., 1858 (Pt. 2), pp. 6, 7.

271. On Vision through the Foramen Centrale of the Retina.—Brit. Assoc. Rep., 1858 (Pt. 2), p. 7.

272. On certain Abnormal Structures in the Crystalline Lenses of Animals, and in the Human Crystalline.—Brit. Assoc. Rep., 1858 (Pt. 2), pp. 7-10.

273. On the Crystalline Lens of the Cuttle-Fish.—Brit. Assoc. Rep., 1858 (Pt. 2), pp. 10-13.

274. On the Use of Amethyst Plates in Experiments on the Polarization of Light.—Brit. Assoc. Rep., 1858 (Pt. 2), p. 13.

275. On the Optics of Photography.—Photogr. Soc. Journ. iv., 1858, pp. 83-89.

276. Account of a new Photographic Process by M. Dupuis.—Photogr. Soc. Journ. iv., 1858, p. 88.

277. On a new Species of Double Refraction.—Brit. Assoc. Rep., 1859 (Pt. 2), pp. 10, 11.

278. On the Decomposed Glass found at Nineveh and other Places.—Brit. Assoc. Rep., 1859 (Pt. 2), p. 11 ; 1860 (Pt. 2), pp. 9-12.

279. On Sir Christopher Wren's Cipher, containing Three Methods of Finding the Longitude.—Brit. Assoc. Rep., 1859 (Pt. 2), pp. 34, 35.

280. Note sur la Polarisation de la Lumière des Comètes.—Paris, Comptes Rendus, xlviii., 1859, pp. 384-385.

281. On the Coloured Houppes or Sectors of Haidinger.—Phil. Mag. xvii., 1859, pp. 323-326.

282. On the Lines of the Solar Spectrum.—Roy. Soc. Proc. x., 1859-60, pp. 339-341.

283. On some Optical Illusions connected with the Inversion of Perspective.—Brit. Assoc. Rep., 1860 (Pt. 2), pp. 7, 8.

284. On Microscopic Vision, and a new Form of Microscope.— Brit. Assoc. Rep., 1860 (Pt. 2), pp. 8, 9.

285. Observations sur un point d'Histoire de l'Optique.—Paris, Comptes Rendus, li., 1860, pp. 425-429, 467.

286. On the Invention of the Stereoscope in the Sixteenth Century, and of Binocular Drawings by Jacopo Chimenti da Empoli.— Photogr. Soc. Jour. vi., 1860, pp. 232, 233.

287. On Binocular Lustre.—Brit. Assoc. Rep., 1861 (Pt. 2), pp. 29-31.

288. On the action of Uncrystallized Films upon Common and Polarized Light.—Edinb. Roy. Soc. Trans. xxii., 1861, pp. 607-610 ; Phil. Mag. xxii., 1861, pp. 269-273.

289. On the Pressure Cavities in Topaz, Beryl, and Diamond, and their Bearing on Geological Theories.—Edinb. Roy. Soc. Trans. xxiii., 1861, pp. 39-44 ; Phil. Mag. xxv., 1863, pp. 174-180.

290. On the Existence of Acari between the Laminæ of Mica in Optical Contact.—Edinb. Roy. Soc. Trans. xxiii., 1861, pp. 95, 96.

291. On certain Vegetable and Mineral Formations in Calcareous Spar.—Edinb. Roy. Soc. Trans. xxiii., 1861, pp. 97, 98.

292. On the Structure and Optical Phenomena of Ancient Decomposed Glass.—Edinb. Roy. Soc. Trans. xxiii., 1861, pp. 193, 204.

293. On the Polarization of Light by Rough and White Surfaces. —Edinb. Roy. Soc. Trans. xxiii., 1861, pp. 205-210 ; Phil. Mag. xxv., 1863, pp. 344-350.

294. Observations on the Polarization of the Atmosphere, made at St. Andrews in 1841, 1842, 1843, 1844, and 1845.—Edinb. Roy. Soc. Trans. xxiii., 1861, pp. 211-240.

295. On certain Affections of the Retina.—Phil. Mag. xxi., 1861, pp. 20-24.

296. Description of the Lithoscope, an Instrument for Distinguishing Precious Stones and other Bodies.—Edinb. Roy. Soc. Trans. xxiii., 1861, pp. 419-425.

297. On the action of various Coloured Bodies on the Spectrum. —Phil. Mag. xxiv., 1862, pp. 441-447.

298. On Photographic and Stereoscopic Portraiture.—Photogr. Soc. Jour. vii., 1862, p. 190.

299. On Discoveries and Inventions in Photography.—Photogr. Soc. Jour. vii., 1862, p. 183.

300. Sir D. B. and J. H. Gladstone. On the Lines of the Solar Spectrum.—Phil. Trans., 1860, pp. 149-160.

301. Sir D. B. and Gordon. Experiments on the Structure and Refractive Power of the Coats and Humours of the Human Eye.— Edin. Phil. Jour. i. 1819, pp. 42-45.

302. Sir D. B. and L. Horner. On an Artificial Substance resembling Shell ; with an account of an examination of the same.—Phil. Trans., 1836, pp. 49-56 ; Phil. Mag. x., 1837, pp. 201-210.

303. Sir D. B. and Seebeck. Sur le Développement des Forces polarisantes par la pression.—Paris, Soc. Philom. Bull, 1816, pp. 49-51.

304. Sir D. B. and Yelin, Eine neur in den Höhlengen von Mineralien eutdeckte Flüssigkeit von Londerbaren physikalischen Eigenschaften.—Gilbert, Annal. lxxiv., 1823, pp. 331-333.

305. On the Cause and Cure of Cataract.—Edinb. Roy. Soc. Trans. xxiv., pp. 11-14.

306. On Hemiopsy or Half-Vision.—Edinb. Roy. Soc. Trans. xxiv., pp. 15-18.

307. On the Bands formed by the Superposition of Paragenic Spectra produced by the Grooved Surfaces of Glass and Steel (1864). —Edinb. Roy. Soc. Trans. xxiv., pp. 221-225 (1865) ; Edinb. Roy. Soc. Trans. xxiv. (Pt. 1), pp. 227-232.

308. On the Colours of the Soap-bubble (1867).—Edinb. Roy. Soc. Trans. xxiv., pp. 491-504.

309. On the Figures of Equilibrium in Liquid Films.—Edinb. Roy. Soc. Trans. xxiv., pp. 505-513.

310. Description of a Double Holophote Apparatus for Lighthouses, and of a Method of Introducing the Electric or other Lights (1867).—Edinb. Roy. Soc. Trans. xxiv., pp. 635-638.

311. On the Motions and Colours upon Films of Alcohol and Volatile Oils and other Fluids (1867).—Edinb. Roy. Soc. Trans. xxiv., pp. 653-656.

312. On the Fairy Stones found in the Elwand Water near Melrose (1866).—Edinb. Roy. Soc. Proc. v., pp. 567-571.

313. On the Vapour Lines in the Spectrum, 1867.—Edinb. Roy. Soc. Proc. vi., pp. 145-147.

314. On the Radiant Spectrum, 1867.—Edinb. Roy. Soc. Proc. vi., pp. 147-149.

315. On the Motion, Equilibrium, and Forms of Liquid Films (1868).—Edinb. Roy. Soc. Trans. xxv. (Pt. 1), pp. 111-118 ; Edinb. Roy. Soc. Proc. vi., pp. 311-313.


The following are among his contributions to the *North British Review* :—

1. Flourens' Eloge Historique de Baron Cuvier.
2. Pascal's Life, Writings, and Discoveries.
3. Lord Rosse's Reflecting Telescopes.
4. Eusèbe Salverte on the Occult Sciences.
5. Vestiges of the Natural History of Creation.
6. Explanations of ditto.
7. Baron Humboldt's Kosmos.
8. Arago's Eloge Historique de Baron Fourier.
9. Murchison's Geology of Russia.
10. Baron Humboldt's Central Asia.
11. The Revelations of Astronomy.
12. Composition of Water.   Watt and Cavendish.
13. Neptune.   Adams and Leverrier.
14. Photography.
15. Sir John Ross's Antarctic Expedition.
16. On the Construction and Use of Microscopes.
17. Sir John Herschel's Astronomical Observations.
18. Mrs. Somerville's Physical Geography.
19. Johnston's Physical Atlas.
20. Rajah Brooke's Residence in Borneo.

21. Britton's Authorship of Junius.
22. Macaulay's History of England.
23. Layard's Nineveh and its Remains.
24. Railway System of Great Britain.
25. Baron Humboldt's Aspect of Nature.
26. Hugh Miller's Footprints of the Creator.
27. Hunt's Poetry of Science.
28. Tubular Bridges.  Stephenson and Fairbairn.
29. British Association.
30. Sir Charles Lyell's Travels in North America.
31. Babbage's Exposition of 1851.
32. Arago's Life of Carnot.
33. Binocular Vision and the Stereoscope.
34. Prize Essays on the Peace Congress.
35. Search for Sir John Franklin.
36. Prince Albert's Industrial College.
37. Grant's Physical Astronomy.
38. History, Properties, and Origin of the Diamond.
39. Layard's Discoveries in Nineveh and Babylon.
40. Grenville Papers.  Junius.
41. Weld's History of the Royal Society of London.
42. Life and Discoveries of François Arago.
43. Of the Plurality of Worlds.
44. Sir Roderick Murchison's Siluria.
45. Sir Henry Holland's Mental Philosophy.
46. The Electric Telegraph.
47. Muirhead's Life of Watt.
48. Peacock's Life of Dr. Young.
49. Paris Exposition and Patent Laws.
50. Wilson on Colour-Blindness.
51. The Weather and its Prognostics.
52. The Microscope and its Revelations.
53. The Sight, and How to See.
54. Dr. Kane's Arctic Explorations.
55. Memoirs of John Dalton.
56. Quatrefage's Rambles of a Naturalist.
57. Maury's Geography of the Sea.
58. Researches into Light.
59. Popular Electricity.

Printed in the United States
By Bookmasters